IAN . OWEN .

CW00499229

Technician Science

W Bolton

Longman Scientific & Technical,
Longman Group UK Limited,
Longman House, Burnt Mill, Harlow,
Essex CM20 2JE, England
and Associated Companies throughout the world.

© Longman Group UK Limited 1988

All rights reserved; no part of this publication may be reproduced,
stored in a retrieval system, or transmitted in any form or by any
means, electronic, mechanical, photocopying, recording, or otherwise,
without either the prior written permission of the Publishers or a
licence permitting restricted copying issued by the Copyright Licensing
Agency Ltd, 90 Tottenham Court Road, London W1P 9HE.

First published 1988
Second impression 1990

British Library Cataloguing in Publication Data

Bolton, W (William), 1933–
    Technician science.
    1. Science
    I. Title
    500   Q158.5

ISBN 0-582-01656-8

Longman Scientific & Technical,
Longman Group UK Limited,
Longman House, Burnt Mill, Harlow,
Essex CM20 2JE, England
and Associated Companies throughout the world.

© Longman Group UK Limited 1988

All rights reserved; no part of this publication may be reproduced,
stored in a retrieval system, or transmitted in any form or by any
means, electronic, mechanical, photocopying, recording, or otherwise
without either the prior written permission of the Publishers or a
licence permitting restricted copying issued by the Copyright Licensing
Agency Ltd, 90 Tottenham Court Road, London W1P 9HE.

First published 1988
Second impression 1992

British Library Cataloguing in Publication Data

Bolton, W. (William), *1933–*
  Technician science.
  1. Science
  I. Title
  500.2'0246   Q160.2

ISBN 0-582-01656-8

Printed in Malaysia by VP

Contents

Preface

This book has been written with the aim of enabling technicians and potential technicians to:

1. acquire the basic language of science relevant to any study of technology,
2. develop an understanding of fundamental science concepts which can provide a base for further studies in science and technology,
3. develop a basic understanding of the concepts and language of chemical reactions, atoms and molecules, elasticity, forces and static equilibrium, fluid pressure, straight line motion, wave motion, energy, work, heat, electrical circuits involving resistance, power, and the heating, chemical and magnetic effects of electricity.
4. develop the ability to handle physical equations, plotting and analysis of graphs and the establishment of relationships between physical quantities,
5. acquire skills in laboratory investigations, involving the isolation of variables.

The book covers the Business and Technical Education Council's unit 'Science' (2863B) in the BTEC First courses. This unit replaces the BTEC unit 'Physical Science I' (U80/682). The unit is common to BTEC courses for technicians or potential technicians in construction, engineering and science. The book is also relevant to students of many CGLI and other craft or technician courses.

The book has been written to meet the needs of students for a text which includes a clear exposition of principles, worked examples on all key aspects and self-assessment questions with guides to answers for all the questions. My prime concern in writing the text has been to help you, the student.

W. Bolton

1 The language of science

Introduction

The work of technicians, whether they be construction, engineering or science technicians, almost invariably tends to involve certain types of activities:

1. communication of technical information,
2. diagnosis (like a doctor trying to find out what is wrong with a patient), involving measuring, inspecting and faultfinding,
3. practical work, perhaps assembly, maintenance, production or servicing types of operation.

All these activities will involve the language and methods of science. Thus the technical information might be a drawing of a building or perhaps an engineering component. It might be data about the pressure and temperature necessary for a chemical reaction, or perhaps the density of steel. Such types of information involve physical quantities, such as length, pressure, temperature, density, and associated with these will be units of measurement such as millimetres, pascals, degrees celsius, kilograms per cubic metre. Diagnosis involves a knowledge of the language of science and also requires an ability to carry out an investigation in what can be termed a scientific manner. It is also likely to require a knowledge of the scientific principles and laws that lie behind the language. The practical work, and also the diagnosis work, can involve practical skills.

This first chapter can be considered to be an introduction to some of the basic language of science, scientific investigation and the communication of technical information. All these aspects will be developed further in later chapters.

Scientific terms

The following are some commonly used terms in science.

1 A scientific *fact* results from experimental observation and can be reproduced if the experiment is repeated.

2 A scientific *law* is a statement that summarizes a range of observations and presents a general idea of importance.

1

3 A scientific *rule* is the name given to a series of working instructions, perhaps to calculate some quantity.

4 A scientific *theory* is an imaginative picture used to relate a number of observations by a single explanation.

5 A scientific *model* is an imaginative picture used to aid discussion of some phenomena.

Physical quantities and units

Without a unit, a value for a physical quantity has no significance. The international scientific community has adopted a system of units, with symbols for them, which is known as the *SI system*. The system is based on a specification of the units of seven basic quantities.

Basic physical quantity	Name of SI unit	Symbol for unit
Mass	kilogram	kg
Length	metre	m
Time	second	s
Electric current	ampere	A
Temperature	kelvin	K
Luminous intensity	candela	cd
Amount of substance	mole	mol

Units of other physical quantities are obtained from these basic units by multiplying or dividing them. Thus, for instance, area is obtained by multiplying two lengths and so has the unit of metre $\times$ metre or square metre or m^2. Volume is obtained by multiplying three lengths and so has the unit of metre $\times$ metre $\times$ metre or cubic metres or m^3. Density is mass divided by volume and so has the unit of kilogram divided by cubic metres or kilogram per cubic metre or kg/m^3 (this can also be written as $kg\ m^{-3}$). These types of units are called derived units, since they have been derived, i.e. obtained, from manipulating the basic units.

In addition to the basic and derived units there are some units referred to as supplementary units. If these had been derived from the basic units they would have ended up with no unit. One, for example, involves a length divided by a length. The supplementary units are the radian (symbol rad) for plane angle and the steradian (symbol sr) for the solid angle.

Definitions of SI basic units

1 *Mass* The kilogram is the mass of the international prototype kilogram preserved at the International Bureau of Weights and Measures in France. This is in the form of a solid cylinder of a platinum-irium alloy.

2 *Length* The metre is the length equal to 1 650 763.73 wavelengths in a vacuum of a specified radiation from an atom of krypton-86.

3 *Time* The second is the duration of 9 192 631 770 periods of a specified radiation from an atom of caesium-133.

4 *Electric current* The ampere is that constant current which maintained in two straight parallel conductors of infinite length, of negligible circular cross-section and placed 1 metre apart in a vacuum, causes each to exert a force of 2×10^{-7} newton on 1 metre length of the other.

5 *Temperature* The kelvin is the fraction 1/273.16 of the thermodynamic temperature of the triple point of water.

6 *Luminous intensity* The candela is the fraction 1/60 of the luminous intensity per square centimetre of a specified radiator.

7 *Amount of substance* The mole is the amount of substance of a system which contains as many elementary units as there are carbon atoms in 0.012 kg of carbon 12.

Unit prefixes

The metre may not always be a convenient size unit of length. Thus, for example, for the distance between two towns the metre can be a rather small unit of length, while for the thickness of a thin sheet of metal the unit is too large. For this reason prefixes can be attached to units to indicate the number by which the unit has been multiplied. Thus the prefix kilo is used to mean multiplying a unit by 1000 and so a kilometre (km) is 1000 metres. Milli means multiplying by 1/1000 and a millimetre (mm) is one thousandth (1/1000) of a metre. The following table gives the prefixes and the factors by which the unit has to be multiplied when they are used.

Prefix	Symbol	Multiplying factor
atto	a	10^{-18}
femto	f	10^{-15}
pico	p	10^{-12}
nano	n	10^{-9}
micro	μ	10^{-6}
milli	m	10^{-3}
kilo	k	10^{3}
mega	M	10^{6}
giga	G	10^{9}
tera	T	10^{12}

The multiplying factors in the above table have been expressed in 'powers of ten'. The index, or power, is the number written at the top right-hand corner of the number 10. Thus 10^3 is 10 to the power (or index) 3. This means that 10^3 is ten multiplied by itself three times.

$$10^3 = 10 \times 10 \times 10$$

10^9 is 10 to the power 9 and is 10 multiplied by itself nine times.

$$10^9 = 10 \times 10 \times 10 \times 10 \times 10 \times 10 \times 10 \times 10 \times 10$$

When the index is negative it means that the factor is 1 divided by 10 multiplied by itself the appropriate number of times. Thus, for example, 10^{-3} is 1 divided by 10 multiplied by itself three times.

$$10^{-3} = \frac{1}{10 \times 10 \times 10} \left(= \frac{1}{10} \times \frac{1}{10} \times \frac{1}{10} \right)$$

10^{-9} is 1 divided by 10 multiplied by itself nine times.

$$10^{-9} = \frac{1}{10 \times 10 \times 10 \times 10 \times 10 \times 10 \times 10 \times 10 \times 10}$$

The table given previously is based on what are termed the 'preferred' prefixes. Such prefixes involve indices which go up in steps of 3. Other prefixes are, however, used on occasions and the following are some common ones.

Prefix	Symbol	Multiplying factor
centi	c	10^{-2}
deci	d	10^{-1}
deca	da	10^1
hecto	h	10^2

Example

Write a length of 0.004 5 m in millimetres.

Solution

$$0.004\ 5 = \frac{4.5}{10 \times 10 \times 10} = 4.5 \times 10^{-3}$$

Hence 0.004 5 m = 4.5 mm

Example

Write a length of 35 200 m in kilometres.

Solution

$$35\ 200 = 35.2 \times 10 \times 10 \times 10 = 35.2 \times 10^3$$

Hence 35 200 m = 35.2 km

Physical equations

In any equation involving physical quantities not only must the numbers on both sides of the equation balance but so also must the units.

Consider the equation

$$\text{density} = \frac{\text{mass}}{\text{volume}}$$

The unit of density must be equivalent to the unit of mass divided by that of volume. Hence, since the unit of mass is the kilogram and that of volume the cubic metre, then the unit of density must be kilograms per cubic metre.

$$\text{Density in } \frac{\text{kg}}{\text{m}^3} = \frac{\text{mass in kg}}{\text{volume in m}^3}$$

The equation would not balance if, say, we had the density in kg/m^3 with the mass in grammes and the volume in cubic metres.

Example

What is the unit of F in the equation $F=ma$ when m has the unit kg and a the unit m/s^2?

Solution

Unit of F = (unit of m) $\times$ (unit of a)

$$= \text{kg} \times \frac{\text{m}}{\text{s}^2}$$

$$= \text{kg m/s}^2 \text{ (or kg m s}^{-2})$$

Graphs

Graphs are useful ways of displaying the results of an experiment when it is required to show how one quantity depends on the value of another. For instance, assume that it is required to show how the distance travelled by an object is related to the time taken. The most common form of graph uses Cartesian or rectangular axes (as shown in Fig. 1.1). The axes are at right angles to each other with the horizontal axis being termed the x-axis and the vertical axis the y-axis. The term origin is used for the point where both y and x have the value zero.

It is usual to choose for the x-axis the quantity which is changing independently of the other quantity being plotted. The quantity chosen for the y-axis is the one that depends on the x-axis quantity.

Plotting a graph

In plotting a graph the following points should be considered.
1 When you know or have some idea of the range of values likely to be covered by the two variables, choose scales for the two axes so that they spread the points being plotted and do not result in all the points being clustered in just one small area of the graph paper. As far as possible the scales should also be chosen so that any line makes an angle between about 30° to 60° with the axes. It is often not

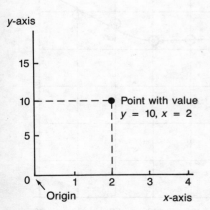

Fig. 1.1 Cartesian axes

necessary for the origin to be on the graph, particularly if this would mean that all the points were clustered in one corner of the graph paper. If the origin is not present, take care in calculating slopes or intercepts.

2 Label the axes with the quantities and their units.

3 Plot the data points, marking each one by either a cross or a dot surrounded by a small circle. The aim of such marking is to make certain that every point is clearly indicated and visible, even when a line is drawn through points.

4 In a graph based on experimentally determined data the points plotted will involve errors made in the measurements and the limited accuracy with which measurements can be made. This results in some scatter of points and thus the 'best' line should be drawn through the data so that there is about the same amount of scatter on one side of the line as the other.

Example

A substance in a test tube was heated and its temperature measured every 10 s. The following results were obtained. Plot a graph of temperature against time.

Temperature in °C	32.0	42.0	49.5	51.5	51.5	53.5	55.0	58.5
Time in s		0	10	20	30	40	50	60	70

Solution

The y-axis scale needs to be chosen to cover the temperature range 32 °C to 58.5 °C. There is no point in including 0 °C on the axis. The x-axis scale needs to go from 0 to 70 s. Figure 1.2 shows the graph and its axes. The axes are labelled with quantity and unit and a smooth line has been drawn which best fits the points. Note that the graph is not a straight line; not all graphs are.

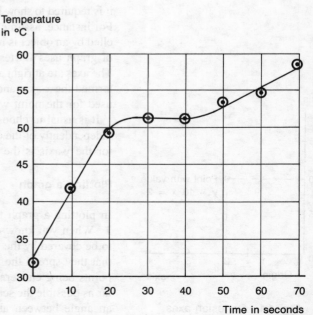

Fig. 1.2 Graph showing the variation of temperature with time for the heated material

Relationships

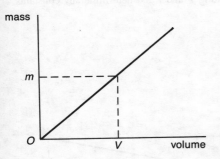

Fig. 1.3 The relationship between mass and volume

For a specific material its mass is proportional to its volume. Thus if the volume is doubled then the mass is doubled; if the volume is trebled then the mass is trebled. We can represent this relationship by

mass $\propto$ volume

with the sign $\propto$ meaning 'is proportional to'.

Another way of writing this relationship is

mass = a constant $\times$ volume

A graph of mass plotted against volume is a straight-line graph passing through the origin (see Fig. 1.3). The slope of the graph is given by

$$\text{slope} = \frac{(V - O)}{(m - O)} = \frac{V}{m}$$

The slope is thus the constant in the equation, i.e., has the same value for any two points on the graph. This constant is given the name density in this particular case. Thus,

mass = density $\times$ volume

The above type of relationship can be expressed in general terms for two variable quantities y and x. If

$y \propto x$

then

$y = mx$

where m is a constant. The graph of y against x gives a straight line passing through the origin with a constant slope equal to m.

Another type of relationship that occurs often in science is where the graph is a straight line but not passing through the origin, as shown in Fig. 1.4. Such a graph can be represented by the equation

$y = mx + c$

where m and c are constants. m is the slope of the graph and c the

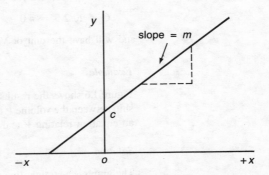

Fig. 1.4 The relationship $y = mx + c$

value of y when x equals zero. This is called the intercept with the y axis. For such a relationship y is not proportional to x.

Example

Figure 1.5 shows the results of an experimental investigation of the relationship between the potential difference V across a resistor and the current I through it. Write an equation relating V and I and determine any constants in the equation.

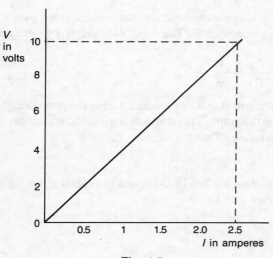

Fig. 1.5

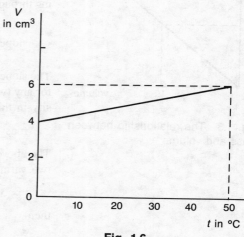

Fig. 1.6

Solution

Since the graph is a straight line passing through the origin then

$$V \propto I$$

and

$$V = CI$$

where C is a constant. C is the slope of the graph, hence

$$C = 10/2.5 = 4.0$$

and will have the unit of V/A.

Example

Figure 1.6 shows the results of an experimental investigation of the relationship between the volume V of a particular substance and temperature t. Write an equation relating V and t and determine any constants in the equation.

Solution

The graph is a straight line not passing through the origin, thus the relationship is of the form $y = mx + c$, i.e.

$$V = mt + c$$

The constant m is the slope of the graph. Thus,

$$m = \frac{(6 - 4)}{(50 - 0)} = 0.04 \text{ cm}^3/°\text{C}$$

The constant is the intercept with the y axis. Thus,

$$c = 4 \text{ cm}^3$$

Hence the equation is

$$V = 0.04t + 4$$

Experimental investigations

In carrying out any experimental investigations there are certain general rules which apply.

1 Be sure you understand what the purpose of the investigation is. It might be, for instance, to obtain a value for some quantity; to determine the relationship between two quantities; to find what, if any, relationships exist between quantities; to investigate a method of making some measurement; to find out what happens in some physical event; etc.

2 Read any instructions given, whether for an experiment or the use of some measuring instrument, and make certain you understand them.

3 Examine any instruments supplied for the investigation and make certain you understand how to use them.

4 Are there any health-and-safety factors which you need to take into account before starting the investigation? If so, make certain you will be able to carry out the investigation without any danger to you, or others or the environment.

5 If instructions are given for the investigation, carry them through in the order they are given. If no instructions are given, plan your own sequence of operations before you start.

6 Record all observations and details of the investigation.

7 Work out results, plot graphs, consider conclusions, etc. during the investigation while you still have time to check a measurement or make some changes which would enable you to obtain a more meaningful conclusion.

Writing reports of experiments

In writing a report of an experimental investigation, while the style to be adopted will depend on who is to read the report, there are a number of aspects which are likely to be common to most reports.

1 The report should have a heading which should clearly and concisely state what the aim of the investigation was.

2 A clear explanation of the methods you used in the investigation, with any problems encountered, should be given. Use diagrams to make your explanation clearer, and probably shorter.

3 A statement of any theoretical basis, perhaps some equation or theory.
4 Details of all observations made.
5 Details of the way you 'process' the observations to perhaps give some value for a quantity or some relationship. This could include a graph.
6 Conclusions. These should give the results of the investigation with a discussion of any limitations on their validity or accuracy.

Self-assessment questions

1 What are the SI units of a) mass, b) length, c) time, d) electric current, e) temperature?

2 Convert the following quantities into the specified units:
a) a length of 0.012 m into mm.
b) a length of 2 560 m into km.
c) a length of 120 mm into m.
d) a current of 0.001 2 A into mA.
e) a current of 160 μA into A.
f) a current of 4 320 A into kA.

3 What is the unit of s in the equation $s=ut$ if u has the unit m/s and t the unit s?

4 What is the unit of p in the equation $p = F/A$ if F has the unit kg m/s^2 and A the unit m^2?

5 A spring was extended by loading with weights and the extension measured. The following are the observations made

Load in grammes	0	50	100	150	200	250
Extension in mm	0	22	40	65	86	109

a) Plot a graph of load (y-axis) against extension (x-axis).
b) Determine the equation relating load and extension, including the value of any constants in the equation.

6 The electrical resistance of different lengths of a wire was measured and the following results obtained.

Resistance in ohms	4.1	8.0	12.2	16.3	20.3	24.6
Length in cm	10	20	30	40	50	60

a) Plot a graph of resistance (y-axis) against length (x-axis).
b) Determine the equation relating resistance and length, including the value of any constants in the equation.

7 The electrical resistance of a length of copper wire was measured at different temperatures and the following results obtained.

Resistance in ohms	24.0	26.9	29.9	32.7	35.7	38.5
Temperature in °C	15	30	45	60	75	90

a) Plot a graph of resistance (y-axis) against temperature (x-axis).
b) Determine the equation relating resistance and temperature, including the value of any constants in the equation.

8 Which of the following will give straight-line graphs? State in each case whether the straight line will pass through the origin, the value of its slope and any intercept value.
a) A graph of potential difference V (y-axis) plotted against current I (x-axis) when $V = IR$.

b) A graph of force F (*y*-axis) plotted against *x* (*x*-axis) where $F = kx$.

c) A graph of distance travelled *s* (*y*-axis) plotted against time *t* (*x*-axis) when $s = \frac{1}{2}at^2$.

d) A graph of the expansion *e* (*y*-axis) of a rod plotted against temperature *t* (*x*-axis) when $e = L\alpha t$.

e) A graph of the force *F* (*y*-axis) needed to break wires of a particular material plotted against their radii *r* (*x*-axis) when $F = \pi r^2 S$.

f) A graph of velocity *v* (*y*-axis) plotted against time *t* (*x*-axis) when $v = u + at$.

2 Chemical reactions

Basic terms

The following are some of the terms frequently used in any discussion of chemical reactions.

1 An *element* is a substance which cannot be decomposed into a simpler substance. Thus hydrogen is an element but common salt is not, since hydrogen cannot be decomposed into anything simpler but common salt can be decomposed to give the elements of sodium and chlorine.

2 An *atom* is the smallest part of an element which can take part in a chemical reaction and which retains the properties of the element.

3 A *compound* is a substance containing two or more elements chemically combined. A compound has properties which are different from the constituent elements. Thus common salt is a compound of sodium and chlorine and has properties vastly different from those of sodium, a highly reactive substance which has to be stored under an inert liquid to stop it reacting with air, and chlorine, a poisonous gas.

4 A *molecule* is the smallest part of a substance which can have a separate stable existence. Thus hydrogen gas exists as molecules each consisting of a combination of two hydrogen atoms. Steam is water in the form of molecules, each molecule consisting of a combination of two hydrogen atoms with one oxygen atom.

5 A *chemical reaction* is the term used to indicate interactions between substances which result in a rearrangement of their atoms. Thus hydrogen can be made to react with oxygen to give water. This involves the initially separate hydrogen and oxygen atoms rearranging themselves so that they combine to give the water molecule. In this reaction two hydrogen atoms combine with each oxygen atom.

6 A *mixture* is a combination of substances without any chemical reaction. Each of the substances brought together in a mixture retains its separate identity and properties.

7 A *solution* is a mixture in which one or more substances have been dissolved in another. Such a mixture is completely homogeneous, i.e. the same composition throughout its volume. The mixture remains homogeneous however long you wait as the substances do not separate out with time. An example of a solution is sugar dissolved in water.

8 A *suspension* is a mixture of a solid and a liquid when the solid does not dissolve. Such a mixture is not homogeneous and the solid separates out from the liquid with time. An example of this is sand in water. In a suspension the solid particles are often visible to the naked eye.

Symbols, formulae and equations

Each element is represented by a symbol, e.g. hydrogen as H, carbon as C, iron as Fe. The symbol is either one or two letters with the first letter a capital letter and the second letter a small letter.

Each compound can be represented by a formula. This gives the symbols of the atoms forming the compound and their numbers in the molecule. Thus, for example, water is H_2O, i.e. two atoms of hydrogen combined with one atom of oxygen.

A chemical equation is a concise way of describing a chemical reaction. For example, copper can react with oxygen to give an oxide of copper. This can be represented by the equation

$$2Cu + O_2 \rightarrow 2CuO$$

In such an equation we are stating that two atoms of copper combined with each molecule of oxygen, each molecule being two oxygen atoms, to give two molecules of copper oxide. The equation balances in that we have the same number of each type of atom on the left hand side as we have on the right.

Left hand side	Right hand side
2 copper atoms	2 copper atoms
2 oxygen atoms	2 oxygen atoms

The states of the reacting substances and products can be included in the equation, being written in brackets after each formula. The symbols used are:

(s) for substances as solids
(l) for substances as liquids
(g) for substances as gases
(aq) for substances as solutions with water.

Thus, our equation above, can be written as

$$2Cu(s) + O_2(g) \rightarrow 2CuO(s)$$

Example

For the following chemical reactions what are the values of x and y?

a) $2Mg + O_2 \rightarrow 2Mg_xO_y$
b) $3Fe + 2O_2 \rightarrow Fe_xO_y$
c) $C + O_2 \rightarrow C_xO_y$

Solution

For each of the equations the number of each type of atom on the left-hand side of the equation must balance that on the right-hand side.

a) For magnesium $2 = 2x$ and so $x = 1$
For oxygen $\quad 2 = 2y$ and so $y = 1$

b) For iron $\quad\quad 3 = x$
For oxygen $\quad 4 = y$

c) For carbon $\quad 1 = x$
For oxygen $\quad 2 = y$

Atomic structure model

The atoms of all elements are made up of three basic particles. These are the *proton*, the *neutron* and the *electron*. The proton is positively charged, the electron is negatively charged, and the neutron carries no charge. The proton and the neutron have essentially the same mass, about 1800 times that of the electron.

The protons and neutrons are packed closely together in a minute nucleus within an atom. Surrounding the nucleus and spread out within the rest of the atom are the electrons. A useful model for the atoms is to consider the electrons orbiting the nucleus rather like the planets orbit the sun. The simplest atom is hydrogen which has one electron orbiting a nucleus which contains just one proton.

An atom has no overall electrical charge. This is because the atom contains equal numbers of protons and electrons, the charge on one proton cancelling out the charge on one electron.

An element is composed of atoms all having the same number of protons. Thus all hydrogen atoms have one proton, all oxygen atoms have eight protons, and all iron atoms have 26 protons. The number of neutrons within atoms of an element can vary. Atoms with the same number of protons but different numbers of neutrons are called isotopes. Thus we can have oxygen existing with:

8 protons and 8 neutrons
or 8 protons and 9 neutrons
or 8 protons and 10 neutrons.

In all cases the number of electrons equals the number of protons and so is 8.

The composition of air

Air is a mixture and its composition can vary. In general, the composition of air by volume is approximately:

Nitrogen	78%
Oxygen	21%
Argon	0.93%
Carbon dioxide	0.03%

plus very small amounts of other gases.

The above is the data for dry air. Air can also contain water vapour, the amount of such vapour varying from place to place, and day to day. At a temperature of about 18 °C the percentage of water vapour by volume can be up to about 4%.

An investigation of combustion

Fig. 2.1 An investigation of combustion

A simple experimental investigation of combusion is that of the burning of a candle. A candle is lit and then covered with a beaker, as shown in Fig. 2.1. What happens?

1. The candle burns for a while and then goes out.
2. A mist forms on the inside of the beaker.

Further tests can be carried out to aid the analysis of the above results.

3. A lighted splint introduced into the beaker, after the candle has gone out, also goes out.
4. If the beaker is lifted and 'fresh' air allowed to enter the beaker the candle will again burn.

A conclusion is that the burning of the candle has removed something from the air — something that is necessary for the burning to occur. This we call oxygen.

5. The mist that forms on the inside of the beaker can be tested. White anhydrous copper sulphate turns blue. This is a test for water.
6. If lime water is introduced, on the end of a glass rod, into the air in the beaker after the candle has gone out it turns milky. This is a test for carbon dioxide.

A conclusion from the above is that when the burning takes place water and carbon dioxide are produced.

This event can be represented by the equation

$$\text{candle(s)} + \text{oxygen(g)} \rightarrow \text{water(g)} + \text{carbon dioxide(g)}$$

The water vapour condenses on the colder parts of the container.

Combustion

When a substance burns in air it combines with the oxygen in the air. Without oxygen the burning cannot occur. A chemical reaction occurs between the substance and the oxygen.

Coal, oil and gas are called fuels. When they burn in air heat is produced. The chemical reaction is

$$\text{fuel} + \text{oxygen} \rightarrow \text{carbon dioxide} + \text{water} \ (+ \text{ heat energy})$$

The reaction of metals with oxygen

Most metals when heated in air gain weight. This is because the metal combines with the oxygen in the air to produce compounds called *oxides*. The term oxide is used for a compound formed during a reaction between an element and oxygen. Thus, for example, if copper is heated in air it loses its shiny metallic copper appearance and in gaining weight becomes black. This is because the copper has taken oxygen out of the air, combined with it and produced black copper oxide (CuO).

$$2Cu(s) + O_2(g) \rightarrow 2CuO(s)$$

Similarly, iron combines with the oxygen in air to give an iron oxide (Fe_3O_4).

$$3Fe(s) + 2O_2(g) \rightarrow Fe_3O_4(s)$$

Not all metals combine with oxygen. Gold and silver do not combine with oxygen, however much they are heated.

The reactions of metals with the oxygen in the air have a considerable effect on the uses we make of the materials. Such reactions do occur at room temperature, but at a slower rate than when they are heated to a high temperature. The following are some examples.

1. Aluminium becomes coated with a thin layer of the oxide in air. This oxide forms a durable protective surface film which prevents further reactions between the aluminium and the oxygen. The result is that aluminium is said to have good corrosion resistance, retaining its metallic appearance.
2. Copper develops a thin layer of oxide on its surface which forms a durable protective surface film which prevents further reactions between the copper and the oxygen. Copper has good corrosion resistance.
3. Gold and silver are not oxidized in air and thus retain their metallic sheen. Gold and silver objects made centuries ago and buried in the ground or ancient Egyptian burial tombs are still unoxidized and retain their metallic surfaces when excavated today.

An investigation of rusting

Under what conditions will iron rust? The following is a simple investigation into rusting. The possible factors that seem worth investigating are:

1. whether air (oxygen) is necessary;
2. whether water is necessary;
3. whether both air and water together are necessary.

To explore these factors three tests can be performed. In the first iron nails are placed in a test tube and, to ensure that only dry air is present with no traces of water, anhydrous calcium chloride which absorbs water is added. The test tube is sealed to prevent any moisture entering the tube. This arrangement then explores the effect of the first factor. The second factor can be explored by completely filling the second test tube containing iron nails with water which has been

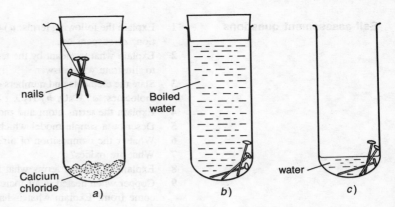

nails

Boiled water

water

Calcium chloride

a)

b)

c)

Fig. 2.2 An investigation of rusting

boiled. The boiling is necessary to remove all the air that is dissolved within water. The test tube is sealed to prevent any air entering. The third test tube is left open and the nails in the test tube rest partly in water and partly in air. Figure 2.2 shows these three arrangements.

The three arrangements are left for a few days. The results are that no rusting occurs when only dry air is present, no rusting occurs when only water is present, but rusting does occur when both air and water are present.

Rusting

Experiment shows that both oxygen and water are necessary for iron to rust. Rusting can be considered to involve the following two chemical reactions:

iron(s) + oxygen(g) → iron oxide(s)
iron oxide(s) + water(l) → hydrated iron oxide(s)

It is the hydrated iron oxide that is rust. Such a layer on the surface of iron offers no protection to the iron, since the rust is very porous and in fact aids the flow of water to the unaffected iron beneath the rust.

A simple method of preventing rust is to coat the iron with some material which will prevent air and water coming into contact with the iron. Such coatings include oil or grease; paint; plastic film; and other more corrosion resistant metals, e.g. zinc (which is known as galvanizing).

Another way of reducing the corrosion of iron is to change the constituent materials in the iron. Iron as a construction material is not pure iron but an alloy involving a mixture of iron with other elements. By choosing appropriate elements, alloys can be produced with much better corrosion resistance, e.g. stainless steel.

A motor car is an example of an iron structure exposed to air and water and where rust can be a serious problem. To overcome this problem the underbody of the car may be undersealed, i.e. coated with a tar-like substance; the bodywork is painted and engine components are covered in oil and grease.

Self-assessment questions

1 Explain the following terms: *a*) element, *b*) compound, *c*) mixture, *d*) solution, *e*) suspension.

2 Explain what is meant by the term 'chemical reaction' and give an example to illustrate your answer.

3 State the elements and numbers of their atoms used to make up the following molecules: *a*) CuO, *b*) H_2O, *c*) Fe_3O_4, *d*) CO_2.

4 Explain the terms atom and molecule.

5 Describe a simple model which illustrates the structure of atoms.

6 What is the composition of air?

7 What are oxides?

8 Explain in general terms what happens when a substance burns in air.

9 Copper when heated in air increases in weight. Where does the extra weight come from? Explain what is happening to the copper.

10 What conditions are necessary for rusting to occur?

11 Give three ways of preventing rusting of iron.

12 Give three examples of where rusting causes damage to components.

13 Determine x and y in the following chemical equations.

a) $S + O_2 \rightarrow S_xO_y$

b) $2Zn + O_2 \rightarrow 2Zn_xO_y$

c) $2Na + O_2 \rightarrow Na_xO_y$

3 Statics

Basic terms

The following are basic terms associated with any discussion of statics.

1 The term *statics* is used to describe the study of forces acting on rigid bodies when the bodies do not move and so remain at rest.

2 The term *at rest* is used to signify that a body does not change its position with respect to objects in its immediate surroundings and the observer who is stating that the object is at rest.

3 A *force* is used to describe an action which changes, or if not counterbalanced would change, the state of rest of a body on which it acts.

4 The *newton* (N) is the SI unit of force.

5 All objects on the Earth's surface, or in the vicinity of the Earth, experience a gravitational force due to its mass. At the surface of the Earth this force is often called the *weight* of an object, where

$$\text{weight in newtons} = mg$$

with m being the mass in kilograms and g the acceleration due to gravity in m/s^2. Approximately this gives

$$\text{weight in newtons} = m \times 9.8$$

Thus an object with a mass of 3 kg will have a weight at the Earth's surface of about $9.8 \times 3 = 29.4$ N.

6 The *mass* of an object is a measure of the quantity of matter in the object.

7 The *kilogram* (kg) is the SI unit of mass.

Structures

The term *structure* covers a wide range of objects. For example, it covers bridges, buildings, cranes, boats, trains, machines, etc. All involve the designer in a consideration of:

1. what forces are acting on each part, and
2. what will happen to the materials as a result of the forces?

This chapter introduces you to a consideration of forces acting on simple objects and the effects of the forces on the materials.

19

Tension and compression

A length of material is said to be in *tension* when the forces acting on it are trying to stretch it and make it longer (Fig. 3.1(*a*)). A length of material is said to be in *compression* when the forces acting on it are trying to squash it and make it shorter (Fig. 3.1(*b*)).

The arrows indicate the direction of the forces

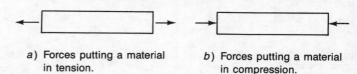

Fig. 3.1

a) Forces putting a material in tension.

b) Forces putting a material in compression.

A part of a structure which is in tension is called a *tie*, a part that is in compression a *strut*. If a tie is cut the two ends move apart and a gap occurs, if a strut is cut no gap occurs since the two ends are pressed together. Figure 3.2 shows two examples of ties and struts.

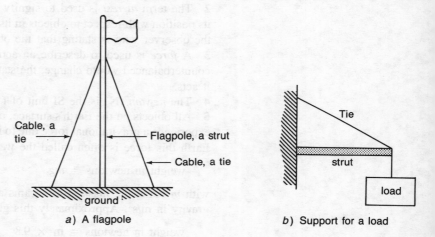

Fig. 3.2

a) A flagpole

b) Support for a load

Hooke's law

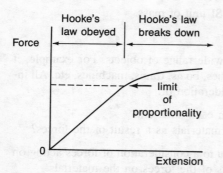

Fig. 3.3 A force—extension graph

A material is said to obey *Hooke's law* if, when it is stretched, the extension is directly proportional to the applied force. This means that to double the extension the force must be doubled, to treble the extension the force must be trebled. A graph of the force plotted against the extension is a straight line passing through the origin. Hence the relationship can be described by

$$\text{force } F \propto \text{extension } x$$

Hence

$$F = kx$$

where k is a constant. This constant is often referred to as the force constant.

Many materials are found to reasonably obey Hooke's law, but only up to some limiting value of force. Beyond such force the proportionality breaks down, see Fig. 3.3.

Example

If a force of 200 N causes a length of wire to extend by 1.0 mm, what will be the extension with a force of 300 N if Hooke's law is obeyed?

Solution

The equation $F = kx$ is used, hence for the force of 200 N,

$$200 = k \times 1.0$$

Hence

$$k = 200 \text{ N/mm}$$

For the 300 N force,

$$F = kx$$
$$300 = 200x$$

$$x = \frac{300}{200} = 1.5 \text{ mm}$$

Stretching materials — an investigation

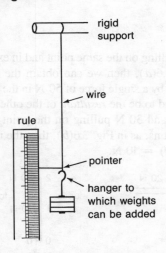

rigid support

wire

rule

pointer

hanger to which weights can be added

Fig. 3.4 Stretching materials, an investigation

Figure 3.4 shows a simple arrangement which can be used to investigate the way materials, such as a wire or a spring, behave when acted on by forces which put the material in tension. The wire is fixed at its upper end to a rigid support and hangs vertically with a weight hanger attached to its lower end. Weight can be added to this hanger so that different forces can be applied to the wire. The applied force is mg, where m is the mass added and g the acceleration of free fall (about 9.8 m/s^2). With m in kilograms and g in metres per second squared the force is in newtons. Attached to the lower end of the suspended wire is a rigid piece of material which acts as a pointer. This moves against a ruler so that changes in the length of the wire being stretched can be measured.

Initially sufficient weights need to be on the hanger for the wire to be taut and hanging vertically. The position of the pointer on the scale is noted. This condition gives us the zero from which all the extensions are measured for future increases in weight. Values are then obtained of the extension when further weights are added and a graph plotted of the force against the extension. If, for the range of forces used and the material, the graph is a straight line passing through the origin then the material is said to obey Hooke's law.

Vectors

The term *vector* is used for any physical quantity for which it is necessary to know both its size and direction in order to determine its effect on some object. Physical quantities for which it is only necessary to know the size are called *scalar* quantities.

Force is a vector quantity. Thus if somebody is pulling on an object we can only work out in which direction the object moves if we know the direction along which the pull is being applied. If we want to know how fast the object will accelerate in that direction we need to know the size of the force as well. Thus both the size and the direction are needed to work out the effect of a force.

Volume is a scalar quantity. A volume only has a size associated with it, we cannot associate any direction with such a quantity.

It is often convenient to represent a vector quantity by a line. The size of the line is drawn to be proportional to the size of the vector and the direction of the line is in the direction of the force, with an arrowhead indicating the direction along the line in which the force is acting. Figure 3.5 shows some examples of such lines drawn to represent forces. Such lines are called vectors.

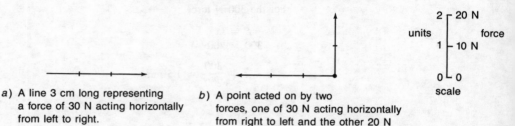

a) A line 3 cm long representing a force of 30 N acting horizontally from left to right.

b) A point acted on by two forces, one of 30 N acting horizontally from right to left and the other 20 N from bottom to top, i.e. vertically.

Fig. 3.5 Vectors, drawn to the scale indicated with 1 unit representing 10 N.

Resultant of two forces

If forces of 20 N and 30 N are pulling on the same point and in exactly the same direction, as in Fig. 3.6(*a*), then we can obtain the same effect by replacing the two forces by a single force of 50 N in the same direction. This single force is said to be the *resultant* of the other two forces. Had the forces of 20 N and 30 N pulling on the point been acting in exactly opposite directions, as in Fig. 3.6(*b*), then the resultant would have been (30 − 20) = 10 N.

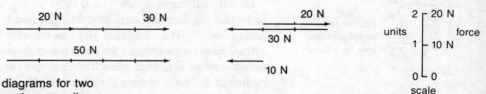

Fig. 3.6 Vector diagrams for two forces acting along the same line, drawn to the scale indicated.

These resultants can be obtained by drawing arrow-headed lines to represent the forces, i.e. the vectors. These can then be combined to give the resultant in what is referred to as a *vector diagram*. For two forces acting along the same line the resultant is obtained by drawing a vector for the first force, then at the arrow end of that vector starting the vector for the second force, taking account of its direction. The resultant is the line drawn between the start of the first vector and the end of the second vector. This method has been used in Fig. 3.6.

The above only involved two forces acting on the same point and along the same line. Now suppose the two forces act on the same point but are inclined to each other at some angle. Figure 3.7(a) shows forces of 30 N and 20 N acting on a point with an angle of 60° between their lines of action. The resultant of these forces can be found by drawing a vector diagram, the diagram being called the *parallelogram of forces*. One procedure for drawing this is as follows, with reference to Fig. 3.7(b):

1. Select a suitable scale for drawing the vectors.
2. Draw a vector to represent the first force, in Fig. 3.7(b) this is the 30 N force represented by line AB.
3. From the start of this first vector, point A, draw the vector to represent the second force of 20 N, i.e. line AD.
4. The parallelogram can be completed by drawing line DC parallel to AB and line BC parallel to AD.
5. The resultant is the vector drawn as the diagonal AC of the parallelogram from point A, the junction of the two vectors. In Fig. 3.7(b) this has a length of about 4.4 units and so is 44 N and a direction of about 23° to the 30 N force.

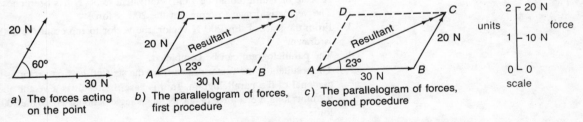

Fig. 3.7 Vectors drawn to the scale indicated, with 1 unit representing 10 N.

a) The forces acting on the point

b) The parallelogram of forces, first procedure

c) The parallelogram of forces, second procedure

An alternative way of drawing the parallelogram, which shows it to be just a version of the triangle of forces (see later in this chapter) is as follows, with reference to Fig. 3.7(c):

1. Select a suitable scale for drawing the vectors.
2. Draw a vector to represent the first force. In Fig. 3.7(c) this is the 30 N force, line AB.
3. From the end of the first vector, point B, draw the vector to represent the second force. In Fig. 3.7(c) this is 20 N, line BC.
4. The parallelograms can be completed by drawing line CD parallel to AB and line DA parallel to BC.
5. The resultant is the vector drawn joining the start of the first vector to the end of the second vector.

The following terms are frequently used when discussing vectors:
1 vectors which all lie in the same plane are said to be *coplanar*,

2 vectors with their lines of action all passing through the same point
are said to be *concurrent*.

Example

Determine the size and direction of the resultant of a force of 200 N and one
of 300 N acting at an angle of 45° to the 200 N force and at the same point.

Solution

The parallelogram of forces can be used to solve this problem, see Fig. 3.8.

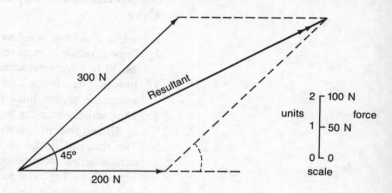

Fig. 3.8 Scale, 1 unit equivalent to 50 N.

1. A scale of 1 unit (could be 1 cm) equivalent to 50 N has been chosen.
2. The vector is drawn to represent the 200 N force.
3. From the start of this 200 N vector, the vector to represent the 300 N
    is drawn.
4. The parallelogram can be completed.
5. The resultant is the vector drawn from the start of the first vector as the
    diagonal of the parallelogram. In this case the result is a vector about
    9.2 units long, i.e. a force of 460 N, at an angle of about 27° to the 200 N
    force.

Resolution of forces into components

Two forces can be replaced by a single force called the resultant using
the parallelogram of forces. The resultant must have the same effect
as the two forces. We can also use the parallelogram in reverse to
replace a single force by two other forces. This operation is said to
be the *resolution* of a force into its components. The combined effect
of the components must be the same as the single force they replaced.

A common form of such a resolution of forces is the resolution into
two components which are mutually perpendicular, i.e. there is 90°
between their lines of action. The procedure for obtaining these components by drawing a vector diagram is as follows, see Fig. 3.9:

1. select a suitable scale for drawing vectors,
2. draw a vector to represent the force being resolved, see Fig. 3.9(*a*),
3. from the start point of this vector draw a line in the direction
    required of one component, shown as component *A* in Fig. 3.9(*b*),

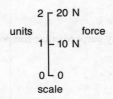

Fig. 3.9 Scale, 1 unit equivalent to 10 N

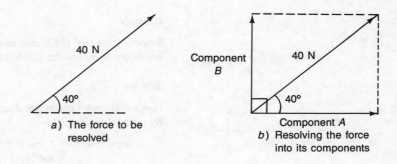

a) The force to be resolved

b) Resolving the force into its components

4. draw a line at right angles to component *A* so that the start of the line is the start of the original 40 N vector, shown as component *B* in Fig. 3.9(*b*),
5. The parallelogram can be completed with lines parallel to components *A* and *B*.
6. The lengths of the components and their directions can be obtained from the parallelogram. *A* has a length of about 3.1 units and thus is 31 N. *B* has a length of about 2.6 units and thus is 26 N.

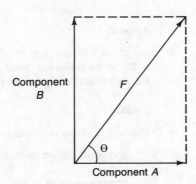

Fig. 3.10 Resolving a force into its components

Instead of drawing a vector diagram to scale to obtain the components they can be obtained fairly easily by calculation. In Fig. 3.10 a force *F* is to be resolved into two right-angled components *A* and *B*, with *A* being at an angle θ to *F*. For the triangle formed by *F* and the two components:

$$\text{since} \quad \cos \theta = \frac{\text{Component } A}{F}$$

$$\text{Component } A = F \cos \theta$$

$$\text{and since} \quad \sin \theta = \frac{\text{Component } B}{F}$$

$$\text{Component } B = F \sin \theta$$

Example

Resolve a force of 120 N into two mutually perpendicular components with one being at 30° to the 120 N force.

Solution

Figure 3.11 shows the vector diagram used to obtain the result by drawing. By calculation

$$\text{component } A = F \cos \theta = 120 \times \cos 30°$$
$$= 104 \text{ N}$$
$$\text{component } B = F \sin \theta = 120 \times \sin 30°$$
$$= 60 \text{ N}$$

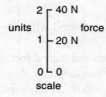

Fig. 3.11 Scale, 1 unit equivalent to 20 N

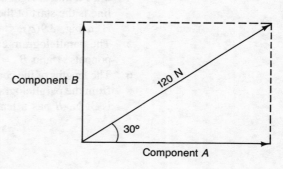

Example

An object with a weight of 100 N rests on an incline which is at an angle of 20° to the horizontal. What are the components of the weight acting down the incline and at right angles to it?

Solution

Figure 3.12 shows the situation. The two components are:

$$\text{component down the incline} = 100 \times \sin 20°$$
$$= 34 \text{ N}$$
$$\text{component at right angles to the incline}$$
$$= 100 \times \cos 20°$$
$$= 94 \text{ N}$$

The results could have been obtained by a scale drawing of the vectors.

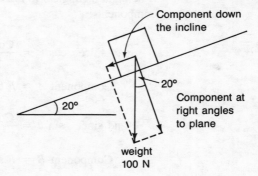

Fig. 3.12

Static equilibrium

When two or more forces acting upon a body are such that the body remains at rest then the forces are said to be in *static equilibrium*. For two forces to be in equilibrium they must be the same size and act along the same straight line but in opposite directions, as shown in Fig. 3.13. For such forces there is zero resultant force, i.e. the overall effect of the two forces is zero since they cancel each other out.

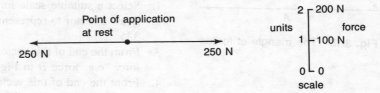

Fig. 3.13 Two forces in static equilibrium

For three forces acting at the same point on a body to be in equilibrium there must be zero resultant force. One way of finding out whether three forces are in equilibrium is to convert the problem to the two-force situation. This involves finding the resultant of two of the forces and then seeing whether this resultant exactly balances the third force. Figure 3.14 shows this being done.

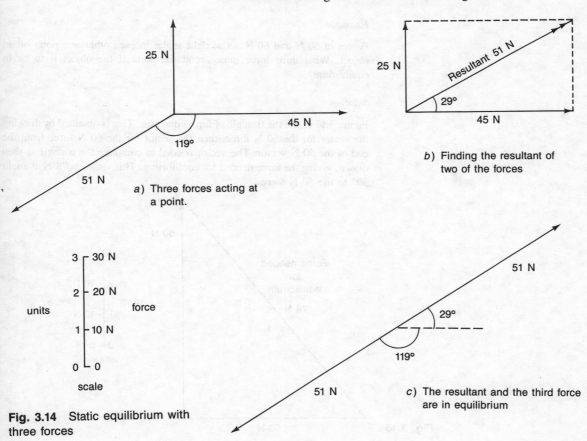

a) Three forces acting at a point.

b) Finding the resultant of two of the forces

c) The resultant and the third force are in equilibrium

Fig. 3.14 Static equilibrium with three forces

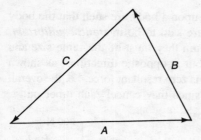

Fig. 3.15 The triangle of forces

There is an alternative way of considering the equilibrium of three forces acting at a point. This is called the *triangle of forces*. If three forces acting at a point can be represented in size and direction by the sides of a triangle taken in order then they are in equilibrium. The following, together with Fig. 3.15, shows how this method of determining equilibrium is used.

1. Select a suitable scale for drawing vectors.
2. Draw a vector to represent one of the forces, e.g. force *A* in Fig. 3.15.
3. From the end of this vector draw a vector to represent the second force, e.g. force *B* in Fig. 3.15.
4. From the end of this vector draw a vector to represent the third force, e.g. force *C* in Fig. 3.15.
5. If the three forces are in equilibrium the end point of the third vector will now be the same as the start point of the first vector, i.e. the resultant shape will be a triangle.

Essentially in drawing the triangle of forces all that has been done is the drawing of the diagram shown in Fig. 3.7(*c*) with the third force being in the opposite direction to the resultant.

Example

Forces of 50 N and 60 N act at right angles to each other at a point on an object. What third force must act at the point if the object is to be in equilibrium?

Solution

Figure 3.16 shows the triangle of forces diagram. This is obtained by drawing the vector for the 50 N force, then the vector for the 60 N force from the end of the 50 N vector. The vector needed to complete the triangle is then drawn, giving the force needed for equilibrium. This vector is 78 N at about 50° to the 50 N force.

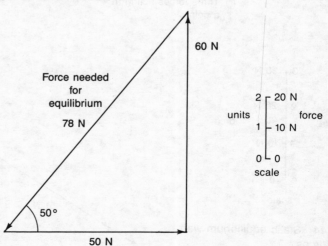

Fig. 3.16

Moments

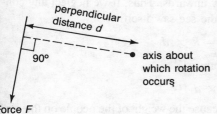

Fig. 3.17 Moment of a force = F d

The *moment of a force* is a measure of its ability to rotate an object about a given axis. The ability of a force to cause rotation depends not only on the size of the force but also on the perpendicular distance of the line of action of the force from the axis of rotation. If you want to open or close a door, i.e. cause the door to rotate about its hinges axis, then you apply a force to the handle which is about the greatest distance from the hinge axis that is possible. If the force is applied closer to the hinge axis it needs to be much larger.

The moment of a force about an axis is the force F multiplied by the perpendicular distance d between the line of action of the force and the axis, see Fig. 3.17.

$$\text{Moment } M = F \times d$$

Moments have the unit Nm, when the force is in N and the distance in m.

Example

A force of 20 N is applied to a spanner to cause it to rotate and undo a nut. The perpendicular distance of the line of action of the force from the centre of the nut is 120 mm. What is the moment of the force about the axis through the centre of the nut?

Solution

Figure 3.18 illustrates this problem. Since

$$M = F \times d$$

then, moment $M = 20 \times \dfrac{120}{1000}$

$$= 2.4 \text{ Nm}$$

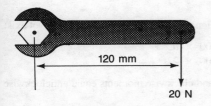

Fig. 3.18

Note that in the above calculation the distance d was converted from millimetres (mm) to metres (m).

Principle of moments

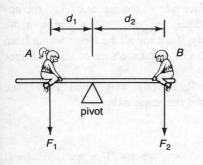

Fig. 3.19 The principle of moments and a see-saw

When a body is in equilibrium the sum of the clockwise moments about an axis must equal the sum of the anticlockwise moments about the same axis. This statement is known as the *principle of moments* and applies to any, and every, axis chosen.

Consider a see-saw, as shown in Fig. 3.19. If the see-saw is balanced, i.e. in equilibrium, with no movement in a clockwise or anticlockwise direction about its pivot, then taking moments about the pivot axis gives

$$\text{clockwise moment} = F_2 \times d_2$$
$$\text{anticlockwise moment} = F_1 \times d_1$$

As there is equilibrium, then

$$F_2 \times d_2 = F_1 \times d_1$$

Suppose moments had been taken about the axis through person A, then:

$$\text{clockwise moment} = F_2 \times (d_1 + d_2)$$

The anticlockwise moment is provided by a reaction force R at the pivot. This force is vertically upwards. Thus, if we neglect any consideration of the weight of the see-saw itself:

$$\text{anticlockwise moment} = R \times d_1$$

Hence, as there is equilibrium:

$$F_2 \times (d_1 + d_2) = R \times d_1$$

The reaction force occurs because the weight of the people on the see-saw is causing the pivot to become squashed. The pivot material is resisting being squashed and pushing back at the see-saw. It is this push which is the reaction force. If you squash a piece of rubber between your fingers you can feel the rubber resisting and pushing back at your fingers.

Example

A person of weight 500 N sits 2.0 m from the pivot axis of a see-saw. At what distance from the pivot should another person of weight 600 N sit if the see-saw is to be balanced?

Solution

Figure 3.20 shows the situation. Hence, taking moments about the pivot axis:

$$\text{clockwise moment} = 600 \times d$$
$$\text{anticlockwise moment} = 500 \times 2.0$$

Hence since there is equilibrium and clockwise moments equal anticlockwise moments:

$$600 \times d = 500 \times 2.0$$
$$d = 1.7 \text{ m}$$

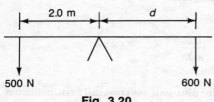

Fig. 3.20

Conditions for equilibrium

The conditions for equilibrium of a body can be summarized as follows:

1. There must be no resultant force in any direction. This means that for a body in equilibrium if we add all the forces in, say, the upward direction they must equal the sum of the forces in the downward direction. Similarly, if we add all the forces acting from left to right then their sum must equal the sum of all the forces acting from right to left.
2. The clockwise moments about any axis must equal the anticlockwise moments about the same axis.

Example

Figure 3.21 shows a plank resting on two supports 2.0 m apart. What will be the reactions at the supports when a person of weight 600 N stands on the plank 0.5 m from one of the supports? Ignore the weight of the plank itself.

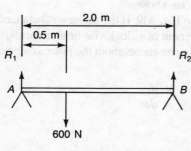

Fig. 3.21

Solution

R_1 and R_2 denote the reactions at the supports. Taking moments about A, gives:

$$\text{clockwise moment} = 600 \times 0.5$$
$$\text{anticlockwise moment} = R_2 \times 2.0$$

Hence, since there is equilibrium:

$$2.0 \times R_2 = 600 \times 0.5$$
$$R_2 = 150 \text{ N}$$

The value of R_1 can be obtained by taking moments about B. An alternative is to use the condition that there will be no resultant force in any one direction. Thus in the vertical direction:

$$\text{upward forces} = R_1 + R_2$$
$$\text{downward forces} = 600$$
$$\text{Hence } R_1 + R_2 = 600$$

Since $R_2 = 150$ N, then

$$R_1 = 600 - 150 = 450 \text{ N}$$

Moments — an investigation

The conditions for equilibrium can be investigated using a ruler or other thin uniform strip of material. The ruler is placed with its centre across a pivot, possibly a triangular piece of wood. Weights can be placed on the ruler and the conditions for equilibrium, i.e. the balance condition, explored. Figure 3.22 shows the apparatus.

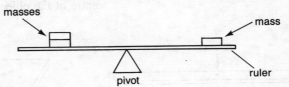

Fig. 3.22 Moments, an investigation

With, say, a mass of 20 g i.e., a weight of $(20/1000) \times 9.8$ N, placed at 20 cm from the pivot on the left-hand side, where should a 10 g mass be placed to obtain balance? Where should a 30 g mass be placed to obtain balance? A table can be compiled as follows:

Left-hand side			Right-hand side		
Mass	Distance from pivot	Anticlockwise moment	Mass	Distance from pivot	Clockwise moment

At balance, how do the anticlockwise moments compare with the clockwise moments?

Centre of gravity

An object, however complex, can be considered to have a single point through which the line of action of the weight of the object can be considered to pass however the body is orientated. This point is called the *centre of gravity*.

If an object is suspended from any point and allowed to freely move to an equilibrium position, the centre of gravity is always directly below the point of suspension. This is the only condition for which the weight gives no moment which would cause the object to rotate, see Fig. 3.23.

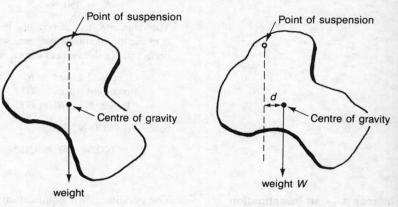

Fig. 3.23 Centre of gravity

No moment due to weight

Clockwise moment of *Wd*

The centre of gravity of homogeneous objects such as a plank, or a rectangular or circular sheet of uniform thickness metal, is at the centre of the object, see Fig. 3.24.

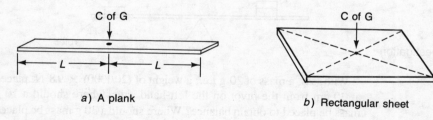

a) A plank

b) Rectangular sheet

c) Circular sheet

Fig. 3.24 In all the above cases, where the material is homogeneous and the thickness is constant, the centre of gravity is halfway through the thickness

Example

A uniform plank of wood of length 3.0 m and mass 10 kg rests on two supports. If the supports are 0.2 m from each end, what will be the reactions at the supports when a person of weight 700 N stands on the plank 1.0 m from one end?

Solution

Figure 3.25 shows the arrangement. The weight of the plank is $10 \times 9.8 = 98$ N and can be considered to act at the centre of the plank. Hence, taking moments about A:

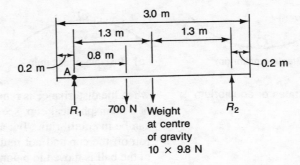

Fig. 3.25

Clockwise moment $= 700 \times 0.8 + 98 \times 1.3$
Anticlockwise moment $= R_2 \times 2.6$

Hence since the anticlockwise moment equals the clockwise moment,

$$R_2 \times 2.6 = 700 \times 0.8 + 98 \times 1.3$$
$$R_2 = 264 \text{ N}$$

Since upward forces must equal downward forces,

$$R_1 + R_2 = 700 + 10 \times 9.8$$
Hence $\quad R_1 = 700 + 10 \times 9.8 - 264$
$$= 534 \text{ N}$$

Centre of gravity — an investigation

The centre of gravity of an object can be experimentally determined by suspending it from a point and allowing it to freely reach equilibrium. When it does, the centre of gravity will be on a vertical line drawn through the point of suspension. If this is repeated for a number of different points of suspension and vertical lines drawn on the object for each equilibrium position, then they will all intersect at one common point — the centre of gravity.

Stability of equilibrium

An object is in equilibrium when the forces acting on the object are such that there is no tendency for the object to move, either by motion in some direction or rotation. The state of equilibrium of an object can, however, be considered to fall into one of three categories of equilibrium.

1 An object is in *stable equilibrium* if when it is slightly disturbed from its rest position it returns to its original position when the disturbing force is removed. An example of this is a ball in a hemispherical cup, see Fig. 3.26(*a*). The ball is in equilibrium when resting at the lowest point in the cup. Disturb it by pushing it away from the centre. When the disturbance is removed the ball returns to its original equilibrium position. This is because the disturbance raises the centre of gravity above its equilibrium position and the resulting moment causes the ball to return to the equilibrium position.

2 An object is in *unstable equilibrium* if when it is disturbed the object moves away from its equilibrium position and does not return

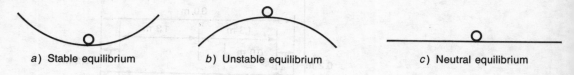

a) Stable equilibrium   *b*) Unstable equilibrium   *c*) Neutral equilibrium

Fig. 3.26 States of equilibrium

when the disturbance is removed. An example of this is a ball on top of a hemispherical cup, see Fig. 3.26(*b*). On top of the cup the ball can be in equilibrium, but a slight disturbance will cause the ball to roll off the cup and not return. This is because the centre of gravity of the ball is moved to below the equilibrium position by the disturbance and the resulting moment does not act in such a direction as to restore the ball to its original position.

3 An object is in *neutral equilibrium* if when it is disturbed the object moves away from its initial position and remains in equilibrium in some new position when the disturbance is removed. An example of this is a ball resting on a horizontal surface, see Fig. 3.26(*c*). In such a case the disturbance does not change the height of the centre of gravity from its height in the initial position.

Toppling over

For an object to be in equilibrium when resting on a surface, the vertical line passing through the centre of gravity of the object must pass within the boundary of contact of the base of the object with the surface. If it does not the object will topple over. This is because the boundary can act as a pivot and if the vertical line through the centre of gravity is outside the base of the object, a moment is produced which

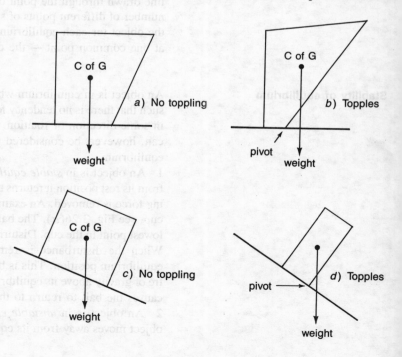

Fig. 3.27 Toppling

is not balanced and so the object rotates about the pivot and topples over, see Fig. 3.27.

Example

A uniform cube of size 10 cm rests on an incline. What will the angle of the incline to the horizontal need to be for the cube to just start to topple? You may ignore any considerations of the cube sliding down the incline.

Solution

Figure 3.28 illustrates the situation when any increase in the angle of the plane will cause the cube to topple. This is when the vertical line through the centre of gravity passes through the base edge. Any greater angle and this line will pass outside the boundary of contact. Since the centre of gravity will be in the centre of the cube, angle θ must be

$$\tan \theta = 5/5 = 1$$
and so $\quad \theta = 45°$

This result applies, regardless of the size of the cube.

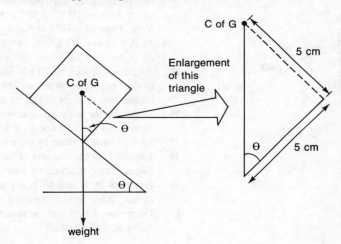

Fig. 3.28 Toppling a cube

Self-assessment questions

1 Explain what is meant by the terms tension and compression when applied to forces acting on a material.
2 What is the difference between a tie and a strut?
3 State Hooke's law and give an example of a situation where it is obeyed.
4 Sketch a force-extension graph for a material that obeys Hooke's law.
5 Complete the missing figures in the following table for the extensions produced by tensile forces if the material obeys Hooke's law.

Force in newtons	0	50	100	150	200
Extension in mm	0	0.2	?	?	?

6 If a force of 10 kN applied to a length of steel causes it to extend by 0.90 mm, what will be the extension when the force is 15 kN if Hooke's law is obeyed?

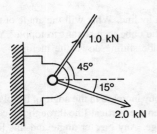

Fig. 3.29

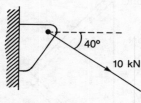

Fig. 3.30

7 A spring has length of 250 mm when unloaded. When a load of 20 N is applied its length becomes 280 mm. What will be its length with a load of 30 N if Hooke's law is obeyed?

8 A wire is stretched by 1.2 mm with a force of 200 N. What force is required to stretch it 2.0 mm if Hooke's law is obeyed?

9 Explain the difference between a vector and a scalar quantity.

10 Determine the size and direction of the resultant for each of the following situations involving two forces acting at a point.
a) A force of 100 N is at right angles to a force of 150 N.
b) A force of 20 N acting along a line at 30° to a force of 50 N.
c) A force of 100 N acting along a line at 60° to a force of 40 N.

11 Explain how the parallelogram of forces can be used to determine a resultant force.

12 Determine the size and direction of the resultant of the two forces acting on the wall bracket shown in Fig. 3.29.

13 A ship is towed by two tugs, the tension in each of the towing cables being 25 N. If each of the cables makes an angle of 20° with the direction of motion of the ship, what is the resultant force on the ship?

14 Resolve the following forces into two mutually perpendicular components with one being in the horizontal direction.
a) A force of 50 N at 30° to the horizontal.
b) A force of 300 N at 50° to the horizontal.
c) A force of 4.2 kN at 45° to the horizontal.
d) A force of 6.0 kN at 20° to the horizontal.

15 An object of weight 2.0 kN rests on an incline which is at an angle of 15° to the horizontal. What are the components of the weight acting down the incline and at right angles to it?

16 Determine the horizontal and vertical components of force acting on the bracket shown in Fig. 3.30 due to a force of 10 kN acting along the cable.

17 Explain what is meant by the term static equilibrium when applied to an object.

18 Explain how the triangle of forces can be used to determine whether three forces acting at a point on an object are in equilibrium.

19 Forces of 20 N and 40 N act at right angles to each other at a point on an object. What third force must act at the point if it is to be in equilibrium?

20 Determine, for each of the situations shown in Fig. 3.31, the third force needed for static equilibrium.

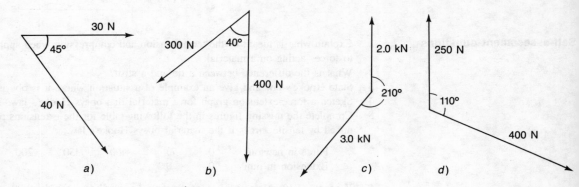

Fig. 3.31

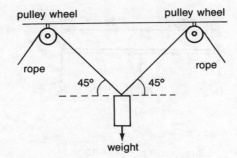

Fig. 3.32

21 An object is being lifted by ropes passing over two pulley wheels, as shown in Fig. 3.32. What will be the tension in each rope when the load being lifted has a mass of 100 kg?

22 Explain the term moment of a force.

23 Calculate the moment of the force about the points A in each of the cases described by Fig. 3.33.

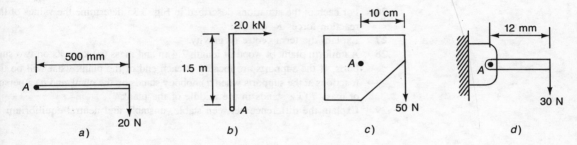

Fig. 3.33

24 A person of weight 600 N sits 2.0 m from the pivot axis of a see-saw. At what distance from the pivot should another person of weight 800 N sit if the see-saw is to be balanced?

25 For each of the situations described in Fig. 3.34 determine the value of the force F needed to obtain equilibrium.

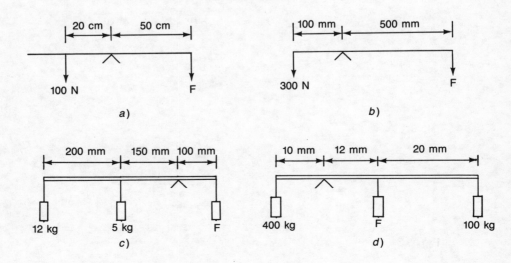

Fig. 3.34

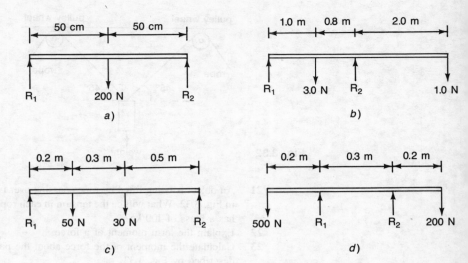

Fig. 3.35

26 For each of the situations described in Fig. 3.35 determine the values of the reaction forces R_1 and R_2.
27 Explain the term centre of gravity.
28 A uniform plank of wood of length 2.4 m and mass 6.0 kg rests on two supports. If the supports are located at each end of the plank what will be the reactions at the supports when a) nobody stands on the plank and b) a person of mass 70 kg stands in the middle of the plank?
29 Explain the difference between stable, unstable and neutral equilibrium.

4 Fluids at rest

Basic terms

The following are basic terms associated with any discussion of fluids.

1 The term *fluid* is used to describe both liquids and gases since both can flow.

2 *Density* ρ is defined as the mass m of a substance divided by its volume V, i.e.

$$\rho = \frac{m}{V}$$

and has the SI unit of kg/m³ when m is in kg and V in m³.

3 *Relative density* is defined by

$$\text{relative density} = \frac{\text{density of substance}}{\text{density of water}}$$

It has no units since it is a ratio.

4 *Pressure p* is defined as force per unit area, ie.

$$p = \frac{\text{force}}{\text{area}}$$

If a force F acts at right angles to an area A then the average pressure over that area is F/A. The SI unit of pressure is the pascal (Pa) and this is when the force is in N and the area in m². Thus

$$1 \text{ Pa} = 1 \text{ N/m}^2$$

The definitions of density, relative density and pressure apply to solids, liquids and gases.

Fluid pressure — an investigation

The following are simple experiments in which no measurements are made but deductions can be made from observing the effects that occur. The first experiment involves half-filling a small polythene bag with water and gripping it tightly around the neck so that the water in the

39

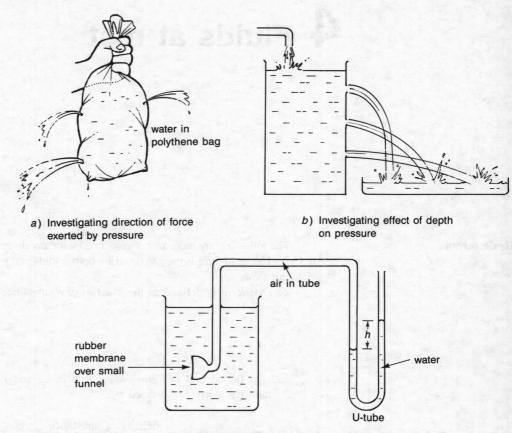

a) Investigating direction of force exerted by pressure

b) Investigating effect of depth on pressure

c) Investigating effect of direction on pressure

Fig. 4.1 Investigating fluid pressure

bag is subject to an increase in pressure, see Fig. 4.1(*a*). A few small pin holes are then put in the bag. What conclusions can be reached about the direction or directions in which the water exerts a pressure on the walls of the polythene bag? Observe the ways in which the water squirts out of the bag. Did you notice that it comes out at right angles to the bag?

The second experiment involves water running into a container to maintain a reasonably constant level of the water in the container, though the container has holes out of which water squirts, see Fig. 4.1(*b*). From observation of the paths of the water emerging from the holes what can be deduced about the effect of water depth on pressure? Did you notice that it squirts out further the greater the depth of water.

The third experiment involves a pressure detecting device. This consists of a piece of rubber stretched over the mouth of a funnel and connected to a U-tube containing water, see Fig. 4.1(*c*). The funnel is kept at the same depth in the liquid but rotated to point in different directions. What happens to the difference in height *h* of the water in the U-tube? What can be deduced about the effect of direction on pressure in a fluid?

The above three experiments lead to the following conclusions:

1. The pressure in a fluid always results in forces at right angles to the surface of its container.
2. The pressure due to a fluid increases as the depth of the fluid increases.
3. The pressure at a given depth in a fluid is the same in all directions.

Pressure variation with depth

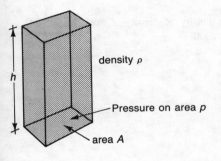

Fig. 4.2

Consider the pressure in a fluid due to a height h of fluid above it, as shown in Fig. 4.2. To determine the pressure we need to find the force acting at right angles to an area A.

Volume of fluid above area $= hA$

Let the density of the fluid be ρ, then since density $=$ mass/volume:

mass of fluid above area $= hA\rho$
weight of fluid above area $= hA\rho g$

where g is the acceleration due to gravity. But the weight of the fluid is the force acting on area A.

Hence force F acting at right angles to area $= hA\rho g$

Since the pressure p acting over the area is F/A,

$$p = \frac{hA\rho g}{A}$$

i.e., $p = h\rho g$

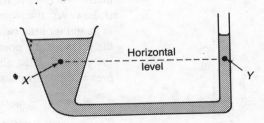

Fig. 4.3

The pressure thus depends on the density of the fluid and the height of fluid above the area. It does not depend on the size of the area considered or on the shape of the column of fluid above. Thus for the arrangement shown in Fig. 4.3, the pressure is the same at X as it is at Y because the height of the fluid above each point is the same.

Example

Calculate the pressure due to water at a depth of 20 mm below a water surface. Water has a density of 1000 kg/m³.

Solution

$$p = h\rho g$$
$$= \frac{20}{1000} \times 1000 \times 9.8$$
$$= 196 \text{ Pa}$$

Example

Calculate the pressure at the foot of a column of mercury of height 760 mm and relative density 13.6 due solely to the mercury. Density of water = 1000 kg/m³.

Solution

Density of mercury = relative density × density of water
$$= 13.6 \times 1000$$

Since
$$p = h\rho g$$
$$p = \frac{760}{1000} \times 13.6 \times 1000 \times 9.8$$
$$= 1.01 \times 10^5 \text{ Pa}$$
$$= 101 \text{ kPa}$$

The U-tube manometer

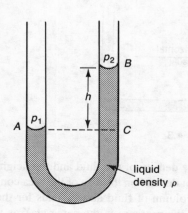

Fig. 4.4 U-tube manometer

Figure 4.4 shows a piece of apparatus that can be used to measure pressure. It uses a U-tube containing a liquid, this often being either water or mercury. The arrangement is called a *manometer*.

When the pressures in the gases above points *A* and *B* are the same then the heights of the columns of liquids in each limb are the same. If, however, the pressures are different then there will be a difference in height *h* of the two liquid levels. When the pressure at *A* is p_1 then the pressure at a point at the same horizontal level in the other limb, i.e. point *C*, must also be p_1. But pressure at *C* = pressure at *B* + pressure due to column of liquid, i.e.,

$$p_1 = p_2 + h\rho g$$

The pressure p_1 at *A* is thus greater than the pressure p_2 at *B* by an amount equal to $h\rho g$. It follows that:

$$p_1 - p_2 = h\rho g$$

Example

What is the pressure difference between the gases above the two limbs of a U-tube manometer containing water of density 1000 kg/m³ when the water in one limb is 10 cm higher than in the other limb?

Solution

Pressure difference = $h\rho g$

$$= \frac{10}{100} \times 1000 \times 9.8$$

$$= 980 \text{ Pa}$$

Atmospheric pressure

If a vacuum is produced above the liquid in one limb of a U-tube manometer and the other limb is left open to the atmosphere, a difference in liquid levels of about 760 mm occurs when the manometer liquid is mercury, see Fig. 4.5. This means that the atmosphere exerts a pressure which is greater than that of a vacuum. Since pressure difference = $h\rho g$, and the density of mercury is 13 600 kg/m³, then

$$\text{pressure difference} = \frac{760}{1000} \times 13\ 600 \times 9.8$$

$$= 1.01 \times 10^5 \text{ Pa}$$

$$= 101 \text{ kPa}$$

Since the pressure of a vacuum is taken to be zero, then the atmospheric pressure is 101 kPa.

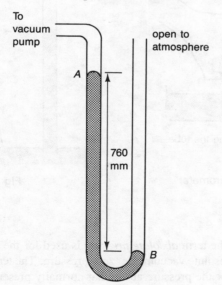

To vacuum pump

open to atmosphere

A

760 mm

B

Fig. 4.5

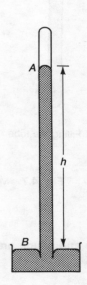

A

h

B

Fig. 4.6 A simple barometer

Simple barometer

The atmospheric pressure can be measured using a version of the U-tube manometer described in Fig. 4.5. Since the limb of the U-tube open to the atmosphere plays no part in the measurement, it can be dispensed with and the arrangement shown in Fig. 4.6 used. This is just a single vertical tube dipping in mercury. The pressure difference is determined between the vacuum at *A* and the atmospheric pressure at *B*, see Fig. 4.6. This arrangement is called a *barometer*.

The vacuum above the mercury in the tube is produced by first filling the tube completely with mercury and then inverting it and standing it vertically in a container of mercury without letting air get into the tube, as shown in Fig. 4.7. The mercury level in the tube then drops to a height which just balances the atmospheric pressure.

An important point to note is that it is the vertical height of the mercury level in the tube above that in the container that matters. If the tube is tilted so that it is not vertical the mercury level in the tube moves to maintain the same vertical height, see Fig. 4.8.

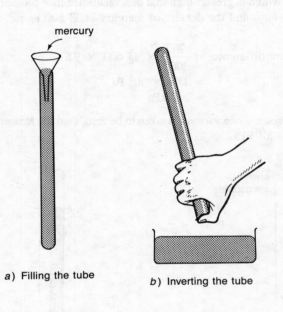

mercury

a) Filling the tube   b) Inverting the tube

Fig. 4.7 Making a simple barometer

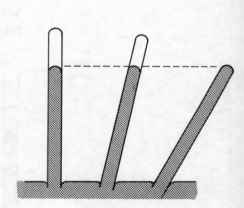

Fig. 4.8

Absolute and gauge pressure

The term *absolute pressure* is used for the pressure above that of an absolute vacuum, i.e. zero pressure. The term *gauge pressure* is used for the pressure above that normally present due to the atmosphere. Thus

    absolute pressure = gauge pressure + atmospheric pressure

Thus when a U-tube manometer is used to measure gas pressure, one limb is connected to the gas whose pressure is being measured, while the other is likely to be open to the atmosphere. This means that the gas pressure is being measured relative to the atmospheric pressure and so the result is the gauge pressure.

Transmission of pressure

When pressure is applied to an enclosed liquid the pressure is transmitted without change to every part of the liquid. This is known as *Pascal's principle*. Thus for the enclosed liquid shown in Fig. 4.9, when the piston is moved in to cause an increase in pressure of the liquid, all parts of the liquid experience the same pressure change.

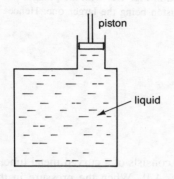

Fig. 4.9

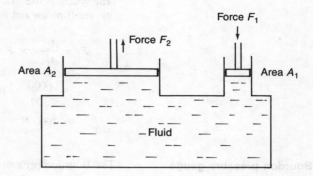

Fig. 4.10

This principle of the transmissibility of pressure through a fluid is the basic principle behind the operation of many hydraulic systems, e.g. the hydraulic brake systems of cars, hydraulic jacks and hydraulic presses.

Figure 4.10 shows a basic hydraulic system. When a force F_1 is applied to the piston moving in the small cylinder a pressure change is produced which is transmitted throughout the fluid.

$$\text{Pressure change} = \frac{F_1}{A_1}$$

where A_1 is the cross-sectional area of the small piston. The larger piston experiences the same pressure change and hence a force acting on it. If this force is F_2 and the cross-sectional area of the larger piston is A_2, then since the pressure change is the same for both pistons, we must have,

$$\frac{F_2}{A_2} = \frac{F_1}{A_1}$$

$$F_2 = \frac{A_2}{A_1} \times F_1$$

Thus if A_2 was ten times greater than A_1 the force acting on the larger piston would be ten times larger than the force used to produce the pressure change. The system acts as a 'force multiplier'.

Example

A hydraulic jack has plunger piston of cross-sectional area of 300 mm^2 to generate pressure changes in an hydraulic fluid. If the pressure acts on a ram piston of cross-sectional area 6000 mm^2, what load on the ram can be lifted when a force of 100 N is applied to the plunger?

Solution

The system is like that described in Fig. 4.10 with the plunger piston being the small piston and the ram piston being the larger one. Hence

$$F_2 = \frac{A_2}{A_1} \times F_1$$

$$= \frac{6000}{300} \times 100 \text{ N}$$

$$= 2000 \text{ N}$$

The Bourdon pressure gauge

The Bourdon pressure gauge consists of a curved metal tube, sealed at one end, as shown in Fig. 4.11. When the pressure in the tube increases the tube tends to straighten out. This causes the sealed end of the tube to move and hence cause a pointer to move across a scale. When the pressure falls the tube curls up again and the pointer then moves in the opposite direction across the scale.

The Bourdon pressure gauge indicates gauge pressure and can be used to measure large pressures. It is a robust instrument and gives a direct reading.

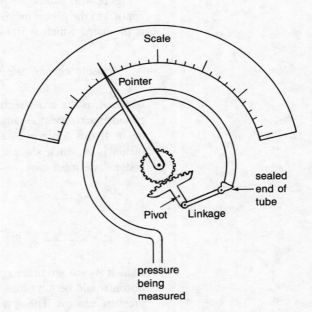

Fig. 4.11 A Bourdon gauge

Archimedes' Principle

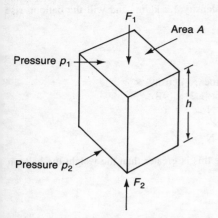

Pressure p_1

F_1

Area A

h

Pressure p_2

F_2

Fig. 4.12 Object immersed in a fluid

Consider an object immersed in a fluid, as shown in Fig. 4.12. The pressure p_2 on the bottom face of the object must be greater than the pressure on the top face since it is at a greater depth in the fluid. Thus

$$p_2 - p_1 = h\rho g \tag{1}$$

where h is the difference in depth in the fluid between the upper and lower faces of the object. Because the pressure is greater on the bottom face it will experience a greater force. Since

$$p = \frac{F}{A}$$

and since for the object the areas of the upper and lower faces are the same, then

$$p_1 = \frac{F_1}{A} \quad \text{and} \quad p_2 = \frac{F_2}{A}$$

Equation (1) becomes

$$\frac{F_2}{A} - \frac{F_1}{A} = h\rho g$$
$$F_2 - F_1 = Ah\rho g$$

Ah is the volume of the solid, V. But $V\rho$ is the mass of the fluid that originally occupied the space now occupied by the object and $V\rho g$ is the weight of this volume of fluid. Thus the difference in the forces between the two faces is

$$F_2 - F_1 = \text{weight of fluid displaced by object}$$

Since F_2 is greater than F_1 we have a net force acting upwards on the object. This is called the upthrust. Hence

$$\text{upthrust} = \text{weight of fluid displaced by object.}$$

This statement is known as *Archimedes' principle* and applies to all objects immersed in fluids, regardless of their shape.

Example

An object of volume 60 cm³ is immersed in a fluid of density 900 kg/m³. What will be the upthrust acting on the object?

Solution

Weight of fluid displaced = volume × density × g

$$= \left(\frac{60}{100 \times 100 \times 100}\right) \times 900 \times 9.8$$
$$= 0.53 \text{ N}$$

Hence the upthrust equals 0.53 N.

Example

A balloon has a volume of 5.0 m³ and a mass of 3.0 kg. What is the upthrust acting on the balloon in air of density 1.2 kg/m³ and will the balloon rise or sink in the air?

Solution

$$\begin{aligned}\text{Weight of air displaced} &= \text{volume} \times \text{density} \times g \\ &= 5.0 \times 1.2 \times 9.8 \\ &= 58.8 \text{ N} \\ \text{Weight of balloon} &= 3.0 \times g \\ &= 29.4 \text{ N}\end{aligned}$$

The upthrust is 58.8 N and since this is greater than the weight the balloon will rise.

Floating

An object floats on a liquid when the upthrust acting on it just balances the weight of the object. When this occurs there is no net force acting on the object and so it does not sink or move upwards. If the weight had been greater than the upthrust the object would sink; if the upthrust had been greater than the weight it would have moved upwards.

When an object is partially immersed in a fluid, e.g. floating with just part of it below the surface, then the fluid displaced is just that which would have occupied the volume of the solid below the surface.

Example

What fraction of the volume of a block of wood of density 600 kg/m³ must be below the surface when it floats in water of density 1000 kg/m³?

Solution

Let V be the volume of the block of wood. Then the weight of the block must be

$$\text{weight of block} = V \times 600 \times g$$

If the fraction of the block below the surface is f then the volume of the block below the surface will be fV. Hence the weight of fluid displaced is $fV \times 1000 \times g$. This is also the upthrust, hence

$$\text{upthrust on block} = fV \times 1000 \times g$$

Since the object floats, the weight of the block must equal the upthrust. Hence

$$fV \times 1000 \times g = V \times 600 \times g$$

$$f = \frac{600}{1000} = \frac{6}{10} = 0.6$$

This is the fraction below the surface.

Hydrometer

Figure 4.13 shows one common form of *hydrometer*. Such instruments are used to measure the density of liquids. The hydrometer consists of a weighted air-filled glass bulb with a long stem containing a density scale. The hydrometer is placed in the liquid whose density is required and floats. The density reading on the scale which is at the level of the liquid in which it floats gives the density of the liquid.

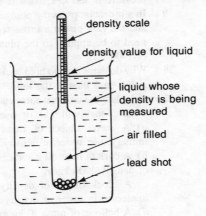

Fig. 4.13 The hydrometer

This instrument relies on the principle of flotation. The depth to which the hydrometer sinks in the liquid depends on the density of the liquid. It floats when the upthrust, i.e. weight of liquid displaced, equals the weight of the hydrometer. The greater the density of the liquid the less the volume of the hydrometer that is below the surface and so the more of the hydrometer that is above the liquid surface.

Self-assessment questions

1 A water tank 4.0 m long and 2.0 m wide, contains water to a depth of 1.5 m. If the density of water is 1000 kg/m^3 what will be *a*) the pressure on the base and *b*) the force acting over the base area due to the water?

2 A cylindrical tank contains oil of relative density 0.8 to a depth of 1.2 m. What will be the pressure on the base of the tank due to the oil? Density of water = 1000 kg/m^3

3 A storage tank contains petrol to a depth of 4.0 m. What is the pressure due to the petrol *a*) on the base of the tank and *b*) 2.0 m up the side wall from the base? Relative density of the petrol = 0.70, density of water = 1000 kg/m^3

4 What is the pressure difference between the gases above the water in the two limbs of a U-tube manometer if there is a difference in water level of 120 mm between them? Density of water = 1000 kg/m^3

5 A simple barometer with mercury as the barometer liquid has a mercury column 750 mm high. What is the atmospheric pressure? The density of mercury is 13 600 kg/m^3

6 What would be the height of the liquid in a simple barometer when the atmospheric pressure is 100 kPa and the liquid used is water? Density of water = 1000 kg/m^3

7 A U-tube manometer is used to measure a gas pressure, one limb being connected to the gas supply and the other left open to the atmosphere. If the manometer liquid, water of density 1000 kg/m³, is 120 mm higher in the limb open to the atmosphere what is the gauge pressure for the gas? Is it higher or lower than the atmospheric pressure?

8 If a manometer indicates a gauge pressure of 20 kPa and the atmospheric pressure is 100 kPa, what is the absolute pressure?

9 In a hydraulic press the plunger piston has a cross-sectional area of 200 mm² and the ram piston a cross-sectional area of 1600 mm². What force would need to be applied to the plunger if the ram is to lift a load with a weight of 1400 N?

10 Why do dams used to hold back water in reservoirs have walls which are thicker at the base than at the top?

11 Calculate the upthrust forces acting on each of the following objects and state whether they will sink or rise in the fluid.

a) A metal object of mass 80 g and density 7000 kg/m³ completely immersed in water of density 1000 kg/m³.

b) A cork of mass 10 g and density 250 kg/m³ completely immersed in water of density 1000 kg/m³.

c) A block of ice of density 920 kg/m³ and volume 10 cm³ completely immersed in water of density 1000 kg/m³.

d) A balloon of mass 10 kg and volume 15 m³ in air of density 1.2 kg/m³.

12 What is the upthrust acting on a boat of mass 300 tonnes (1 tonne = 1000 kg) floating in sea water?

5 Motion

Basic terms

The following are basic terms associated with the discussion of motion.

1 A distinction is drawn between the terms of distance and displacement. *Distance* is the distance along the path followed by an object, whatever the form of the path. *Displacement* is the distance in a straight line between the start and end points of some motion. Figure 5.1 illustrates these two terms. Thus a car might cover distance of 80 km in an hour but its displacement from its start point might be just 5 km in a north-easterly direction. There is no concept of direction associated with distance, so it is a scalar quantity. Displacement always has a direction associated with it and so is a vector quantity.

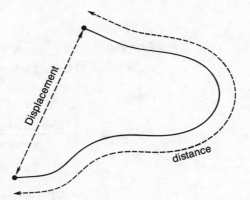

Fig. 5.1 The distinction between displacement and distance

2 A *scalar quantity* requires only a specification of its size for its effects to be determined.

3 A *vector quantity* requires both its size and direction to be specified for its effects to be determined.

4 *Speed* is the rate at which distance is covered. It is a scalar quantity. Thus we can talk of a car having a speed of 80 km/h. This will tell us the distance the car travels in an hour but not where it will be, i.e. its displacement, after an hour.

5 *Average speed* is given by:

$$\frac{\text{distance covered}}{\text{time taken}}$$

During the time interval concerned the car might have a higher or lower speed for a short time but the above equation enables the average speed over the time interval to be determined.

6 *Velocity* is the rate at which displacement increases with time. This is the rate at which distance along a particular straight line is covered with time. It is a vector quantity.

7 *Average velocity* is given by:

$$\frac{\text{displacement}}{\text{time taken}}$$

or

$$\frac{\text{distance covered in a particular direction}}{\text{time taken}}$$

8 A *constant speed* is when equal distances are covered in equal intervals of time, however small we consider the time interval.

9 A *constant velocity* is when equal distances are covered in the same straight-line direction in equal intervals of time, however small the time interval.

Example

A car travels 10 km in 8 minutes. What was its average speed?

Solution

$$\text{Average speed} = \frac{\text{distance covered}}{\text{time taken}}$$

$$= \frac{10}{8}$$

$$= 1.25 \text{ km/minute}$$

If we want the speed in m/s, then

$$\text{average speed} = \frac{10 \times 1000}{8 \times 60}$$

$$= 20.8 \text{ m/s}.$$

Example

A person walks at an average speed of 6.0 km/h for 10 minutes. What distance will be walked?

Solution

$$\text{Average speed} = \frac{\text{distance covered}}{\text{time taken}}$$

$$\text{distance covered} = \text{average speed} \times \text{time taken}$$

$$= 6.0 \times \frac{10}{60}$$

$$= 1.0 \text{ km}$$

Note that the unit of time for the time taken must be the same as the time unit in the speed. Hence, in this case, the time was converted to hours.

Example

A person has a velocity of 0.2 m/s in a north-easterly direction. Where will the person be 200 s after starting?

Solution

$$\text{Average velocity} = \frac{\text{displacement}}{\text{time taken}}$$

$$\text{displacement} = \text{average velocity} \times \text{time}$$

$$= 0.2 \times 200$$

$$= 40 \text{ m in a north-easterly direction}$$

The person will be 40 m along a line in a north-easterly direction from the start point.

Example

What is a speed of 40 km/h in m/s?

Solution

$$1 \text{ km} = 1000 \text{ m}$$
$$1 \text{ h} = 60 \text{ min} = 60 \times 60 \text{ s}$$

Hence

$$40 \text{ km/h} = 40 \times 1000 \text{ m/h}$$

$$= \frac{40 \times 1000}{60 \times 60} \text{ m/s}$$

$$= 11 \text{ m/s}$$

Slope of graphs

The term slope, or gradient, of a graph is used in the same way as the slope or gradient of a hill is referred to. The slope of a hill is given by:

$$\text{slope} = \frac{\text{change in vertical distance}}{\text{change in horizontal distance}}$$

Thus a slope of $\frac{1}{4}$ means that the vertical distance increases by $\frac{1}{4}$ m

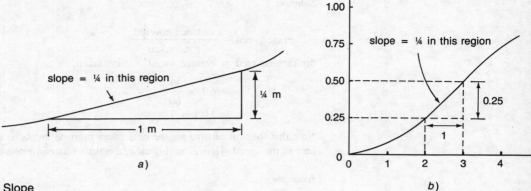

Fig. 5.2 Slope

for every 1 m in the horizontal distance, see Fig. 5.2(*a*). A graph slope of $\frac{1}{4}$ means that the graph line is changing at the rate of $\frac{1}{4}$ of a unit in the vertical, i.e. *y*-axis, direction for every 1 unit change in the horizontal, i.e. *x*-axis, direction, see Fig. 5.2(*b*).

Distance−time graphs

If the distance of an object from some start position is measured for different times from that position, then a distance−time graph can be plotted for the motion. Figure 5.3 shows such a graph. This particular graph has certain features:

1. There is a portion, *AB*, where the distance increases uniformly with time. In the first second the distance increases by 1 m, in the next second it again increases by 1 m. The distance is increasing at the constant rate of 1 m per second. This means that the speed is constant at 1 m/s over the two-second period *AB*. The slope of the graph is constant and is a straight line.

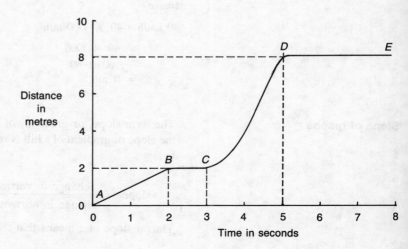

Fig. 5.3 A distance−time graph

2. The portion *BC* has the distance not changing though the time increases from 2 s to 3 s. This can only occur if the object has stopped at the 2 m distance. The slope of the graph is zero and is a horizontal straight line.

3. The portion *CD* has the distance increasing but not at a constant rate, i.e. the slope varies. For a constant rate the graph would have to be a straight line. The average speed over *CD* is the distance travelled (8 − 2 = 6 m) divided by the time taken (5 − 3 = 2 s), i.e.

average speed over *CD* = 6/2 = 3 m/s

4. From *D* to *E* the distance does not change so the object has stopped at the 8 m distance, the slope of the graph is zero.

To calculate the average speed the object takes to go from *A* to *D*:

$$\text{average speed over } AD = \frac{(8 - 0)}{(5 - 0)}$$

$$= 1.6 \text{ m/s}$$

Similarly for the average speed for the object going from *A* to *E*:

$$\text{average speed over } AE = \frac{(8 - 0)}{(8 - 0)}$$
$$= 1.0 \text{ m/s}$$

Thus the average speed for this object obviously depends on what portion of its path we consider. This is because the speed is not constant over the entire path.

The instantaneous speed, i.e. the speed at an instant of time rather than an average over some protracted time interval, is the slope of the graph at the instant concerned. Thus for Fig. 5.4 the slope is:

$$\text{slope of graph at } P = \frac{AB}{BC}$$

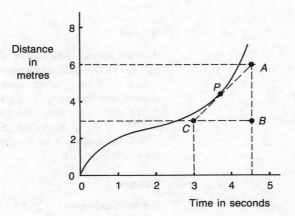

Fig. 5.4 Speed = slope of distance−time graph

$$= \frac{(6 - 3)}{(4.5 - 3)}$$
$$= 2.0 \text{ m/s}$$

In summary: for a distance–time graph:

1. A zero slope means zero speed, the object is stationary.
2. A constant slope means a straight-line graph and a constant speed.
3. A variable slope means a variable speed.
4. The instantaneous speed is the slope of the graph at the instant concerned.

Example

For the graph shown in Fig. 5.5 state whether the object is *a*) at rest, *b*) moving with constant speed, *c*) moving with a varying speed at points *A*, *B*, *C* and *D*.

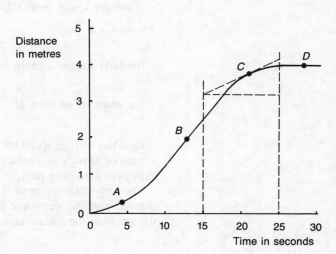

Fig. 5.5

Solution

At *A* the distance is increasing non-uniformly with time, the slope is changing. Hence the speed is changing.

At *B* the distance is increasing uniformly with time, the slope being constant. Hence the speed is constant.

At *C* the distance is increasing non-uniformly with time, the slope is changing. Hence the speed is changing.

At *D* the distance is not increasing with time, the slope being zero. Hence the object is at rest.

Example

For the graph shown in Fig. 5.5 what is the average speed during the first 20 s?

Solution

After 20 s the distance moved is about 3.5 m. Hence the average speed is

$$\frac{3.5}{20} = 0.175 \text{ m/s}$$

Example

For the graph shown in Fig. 5.5 what is the instantaneous speed at 20 s?

Solution

The instantaneous speed is the slope of the graph at the 20 s instant. This is given by:

$$\text{slope} = \frac{1.0}{10} = 0.10 \text{ m/s}$$

Displacement—time graphs

Displacement—time graphs are just like distance—time graphs but the distances are measured in a particular direction along the same straight line. The information that can be obtained from a displacement—time graph is similar to that from a distance—time graph, i.e.

1. An object is at rest when the slope of its displacement—time graph is zero.
2. An object is moving with a constant velocity when the slope of the graph is constant.
3. An object has a changing velocity when the slope of the graph is not constant but changing.
4. The slope of the graph at any instant is the velocity at that instant.

Acceleration

An object is said to be accelerating when its velocity changes. The following are the basic terms.

1 *Acceleration* is defined as the rate of change of velocity with time. The acceleration tells us by how much the velocity is changing in a time interval. It has the SI unit of m/s^2.

2 *Average acceleration* is given by:

$$\frac{\text{change in velocity}}{\text{time interval over which change occurs}}$$

Thus when an object increases its velocity from 2.0 m/s to 5.0 m/s in a time interval of 6.0 s, then

$$\text{average acceleration} = \frac{5.0 - 2.0}{6.0}$$
$$= 0.50 \text{ m/s}^2$$

3 The change in the velocity is the final velocity minus the initial velocity. Thus if an object is increasing its velocity the final velocity is greater than the initial velocity and the acceleration has a positive value. If the final velocity is less than the initial velocity, i.e. the object is slowing down, then the change in velocity is negative and so the acceleration is negative. Negative acceleration is sometimes referred to as a *retardation*. For example, an object had an initial velocity of 4.0 m/s and after 10 s this had dropped to 1.0 m/s. The average acceleration in this case is

$$\frac{(\text{final velocity} - \text{initial velocity})}{\text{time}}$$

$$= \frac{(1.0 - 4.0)}{10}$$

$$= -0.30 \text{ m/s}^2, \text{ the minus sign indicating a retardation.}$$

4 An object is said to have *constant* or *uniform acceleration* if the velocity changes by equal amounts in equal intervals of time, no matter how small a time interval is considered.

Example

A car accelerates from 10 m/s to 12 m/s in 2.0 s. What is its average acceleration?

Solution

$$\text{Average acceleration} = \frac{\text{change in velocity}}{\text{time}}$$

$$= \frac{12 - 10}{2.0}$$

$$= 1.0 \text{ m/s}^2$$

Example

A car travelling at 12 m/s applies its brakes and comes to rest in 4.0 s. What is the retardation?

Solution

$$\text{Average acceleration} = \frac{\text{change in velocity}}{\text{time}}$$

$$= \frac{(0 - 12)}{4.0}$$

$$= -3.0 \text{ m/s}^2$$

Example

A car accelerates uniformly at 1.4 m/s² for 3.0 s. If it had an initial velocity of 8.0 m/s what will be the velocity after the 3.0 s?

Solution

$$\text{Average acceleration} = \frac{\text{change in velocity}}{\text{time}}$$

$$= \frac{\text{(final velocity} - \text{initial velocity)}}{\text{time}}$$

Hence (final velocity − initial velocity) = average acceleration × time

final velocity = average acceleration × time + initial velocity
= 1.4 × 3.0 + 8.0
= 12.2 m/s

Velocity−time graph

Figure 5.6 shows a velocity−time graph for a moving object. This graph has certain features.

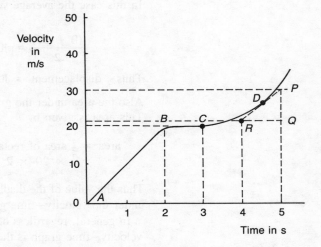

Fig. 5.6 A velocity−time graph

1. Over the portion *AB* the graph has a constant slope. This means that the velocity is changing by equal amounts in equal intervals of time. Thus in the first second it changes from 0 to 10 m/s, in the next second it changes from 10 m/s to 20 m/s. The velocity change is 10 m/s in each second. There is thus uniform acceleration of 10 m/s².
2. Over the portion *BC* the velocity is not changing. This means a constant velocity of 20 m/s.
3. Over the portion *CD* the slope of the graph is continually changing. This means that the velocity is changing by different amounts in equal intervals of time. There is a non-uniform acceleration.

The slope of a velocity−time graph is the acceleration. Thus a constant slope means a constant acceleration. A zero slope means zero acceleration, i.e. constant velocity. The slope of a velocity−time graph at an instant gives the acceleration at that instant. Thus the slope at D is

$$\frac{PQ}{RQ} = \frac{8}{1} = 8 \text{ m/s}^2$$

The area under a velocity−time graph has a significance. Consider the area under the part of the graph AB in Fig. 5.6. The velocity changes at a uniform rate from 0 to 20 m/s in a time interval of 2 s. Since

$$\text{average velocity} = \frac{\text{displacement}}{\text{time}}$$

then   displacement = average velocity × time

The average velocity when the velocity is changing at a uniform rate is $(u + v)/2$, with u being the initial velocity and v the final velocity. In this case the average velocity is

$$\frac{(0 + 20)}{2} = 10 \text{ m/s}$$

Thus   displacement = 10 × 2 = 20 m

Also the area under the graph is the area of triangle under line AB. This area is given by

area = $\frac{1}{2}$ area of rectangle of sides 20 m/s and 2 s
     = $\frac{1}{2}$ × 20 × 2 = 20 m

Thus the value of the displacement obtained is the same as the area under the velocity−time graph.

In general, regardless of the shape of the graph, the area under a velocity−time graph is the displacement covered in the time. If the object is moving in a straight line then it will also be the distance moved along that line. When the area is an irregular shape the area can be estimated by counting the number of graph squares under the line and multiplying the result by the area of one of the graph squares.

Example

For the velocity−time graph in Fig. 5.7 state for the points A, B and C whether the object is moving with a) a constant velocity, b) a uniform acceleration, c) a changing acceleration.

Solution

At A the slope of the graph is changing and so there is a changing acceleration.
   At B the slope of the graph is zero and so there is zero acceleration, i.e. the object is moving with constant velocity.
   At C the slope of the graph is constant and so there is a constant acceleration.

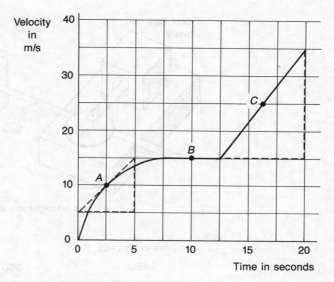

Fig. 5.7

Example

For the velocity–time graph shown in Fig. 5.7 estimate the acceleration at points *A* and *C*.

Solution

Slope of graph *A* is about 10/5
Hence acceleration is about 2.0 m/s^2
Slope of graph at *C* is about 20/7.5
Hence acceleration is about 2.7 m/s^2

Example

For the velocity–time graph shown in Fig. 5.7 estimate the distance travelled in a straight line in the first 20 s.

Solution

The distance travelled is the area under the graph up to the vertical line at 20 s. By counting the squares under the graph there are about $27\frac{1}{2}$ squares. Each square has an area of (5 m/s) × ($2\frac{1}{2}$ s). Hence the total area is

$$27\frac{1}{2} \times 5 \times 2\frac{1}{2}$$
$$= 343.75 \text{ m}$$

The distance travelled is thus about 345 m

Motion — an investigation

One way of investigating the motion of an object in the laboratory is to use ticker tape, i.e. a long narrow strip of paper, and a vibrator which marks the strip every 1/50th of a second. The tape is attached to the object whose movement is to be investigated and its motion pulls the tape through the vibrator. Figure 5.8 shows an arrangement that

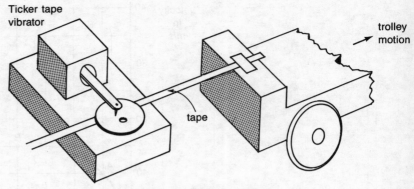

a) The arrangement of the apparatus

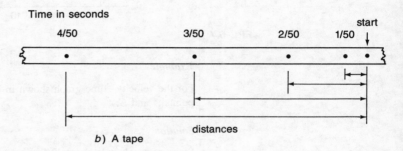

b) A tape

Fig. 5.8

could be used to study the motion of a trolley, perhaps when pulled along a horizontal plane or allowed to roll down an incline.

Distance−time data can be obtained from the ticker tape by measuring the length of the tape from its start point and counting the number of marks made on the tape during the time taken for the object to move that distance. Hence a distance−time graph can be plotted.

Investigate the motion of a trolley rolling down an incline. Try different angles of slope. What type of motion occurs? A distance−time graph for such motion will be of the form shown in Fig. 5.9. The graph is not straight line and so the motion is one of changing velocity. It is not easy to see from such a graph whether the motion is constant acceleration or not. For that, a velocity−time graph is more useful. Such a graph can be obtained from the distance−time graph by obtaining velocity values from the slopes of the distance−time graph for different values of time. An alternative involves working directly from the ticker tape.

The ticker tape is cut into lengths with each corresponding to the same interval of time, say the time taken for 10 dots to be put on the tape. The length of each tape is the distance travelled in that interval. Thus

Fig. 5.9 Distance−time graph for motion down an incline

$$\text{average velocity in the interval} = \frac{\text{length of tape}}{\text{time interval}}$$

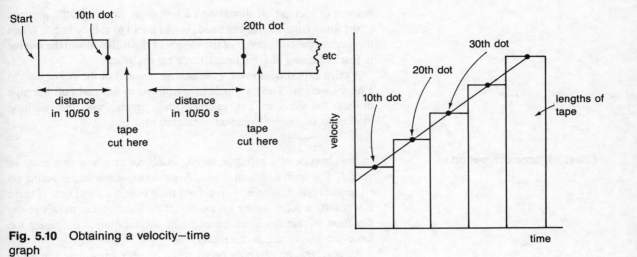

Fig. 5.10 Obtaining a velocity–time graph

Since the time interval is the same for all the cut tapes, then the average velocity is proportional to the length of the tape. The tapes can be mounted side by side on a sheet of paper to give the form of the velocity–time graph, see Fig. 5.10. The resulting graph for motion down an incline is similar to that shown in Fig. 5.10 and indicates uniform acceleration.

Motion down an incline is one of uniform acceleration; the greater the slope of the incline the greater the acceleration.

Another experiment is to pull a trolley along a horizontal plane with a constant force and obtain the velocity–time graph from ticker tape. If a rubber band or elastic thread is stretched and the amount by which it is extended is kept constant, then it is reasonable to suppose that the stretching force is constant. Figure 5.11 shows how such a constant force is applied to a trolley. The result shows that the trolley moves with a constant acceleration under the action of a constant force. (Frictional effects can slightly complicate the result, there being the force from the stretched rubber band and friction both acting on the trolley. The frictional force can be compensated for by initially adjusting the surface on which the trolley rolls so that it is at a slight incline. The

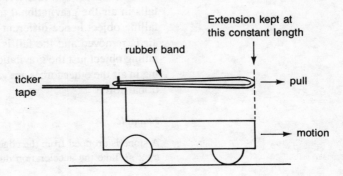

Fig. 5.11 Applying a constant force to a trolley

amount of inclination should be such that the trolley, in the absence of the force from the rubber band, is just about to roll by itself. When this occurs the component of the weight of the trolley down the incline is just balancing the frictional force up the plane.)

A third experiment involves attaching one end of the ticker tape to a heavy weight. The weight is then allowed to fall and pull the tape through the vibrator. The velocity—time graph shows the resulting motion to be one of constant acceleration.

Effect of force on motion

In the absence of a resultant force, an object that is at rest remains at rest. The term resultant is used because there are forces acting on all objects but if there is no resultant their effects cancel out. If there is a resultant force acting on an object then the object moves in the direction of that force with a constant acceleration. The bigger the force the bigger the acceleration.

An acceleration can only occur when there is a resultant force acting on an object. Conversely, when there is no acceleration then there is no resultant force.

Acceleration due to gravity

What is the motion of freely falling objects? The answer to this question would seem to be that it depends on what object is falling when we determine the free-fall motion in air. A feather does not fall in the same way as a large weight. However, if the air is removed, all objects fall freely with a constant acceleration and it is the same acceleration for all objects at the same place, regardless of their weight or shape. This acceleration is called the *acceleration due to gravity* or the *acceleration of free fall*. The value of the acceleration depends on where the objects fall, but at the surface of the earth the value is generally about 9.8 m/s^2.

This acceleration of free fall means we must have a resultant force acting on the falling objects. This force is called the gravitational force. The gravitational force acts on all objects, whether they are falling or not, but the free-fall situation in a vacuum is where this is the only force acting on the object. When an object is at rest, perhaps on the ground, the gravitational force still acts on the object but its effect is cancelled out by other forces acting on the object. When an object falls in air the gravitational force is not the only force acting on the falling object hence different objects fall differently. Only when the air is removed and the fall is in a vacuum is the force acting on the falling object just the gravitational force, though for heavy objects falling in air the other forces are very small in comparison with the gravitational force.

Example

A stone is dropped from the edge of a cliff. What is its velocity after *a*) 1 s, *b*) 2 s? Take the acceleration due to gravity to be 9.8 m/s^2.

Solution

$$\text{Acceleration} = \frac{\text{change in velocity}}{\text{time}}$$

Hence change in velocity = acceleration × time
Initially the stone has zero velocity. Hence

final velocity = acceleration × time

a) At 1 s, velocity = 9.8 × 1 = 9.8 m/s
b) At 2 s, velocity = 9.8 × 2 = 19.6 m/s

Example

A brick falls off the roof of a building. If it falls with the acceleration due to gravity of 9.8 m/s^2:

a) what will be its velocity after 1 s?
b) what will be its average velocity during the 1 s?
c) what will be the distance fallen after 1 s?
d) what will be its velocity after 2 s?
e) what will be the average velocity during the second second?
f) how far will the brick fall in this second second?
g) what will be the total distance fallen after 2 s?

Solution

As with the previous example:

change in velocity = acceleration × time

i.e., final velocity − initial velocity = acceleration × time   (1)

a) Initial velocity at the beginning of the first second is zero, hence

final velocity = 9.8 × 1 = 9.8 m/s

b) $$\text{Average velocity} = \frac{\text{final velocity} + \text{initial velocity}}{2}$$ (2)

$$= \frac{9.8 + 0}{2}$$

$$= 4.9 \text{ m/s}$$

c) Since average velocity $= \dfrac{\text{displacement}}{\text{time}} = \dfrac{\text{distance fallen}}{\text{time}}$

distance fallen = average velocity × time   (3)
= 4.9 × 1
= 4.9 m

d) Using equation (1), after 2 s

final velocity = 9.8 × 2 = 19.6 m/s

e) The velocity at the beginning of the second second is 9.8 m/s and at the end 19.6 m/s. Hence using equation (2)

$$\text{average velocity} = \frac{19.6 = 9.8}{2}$$
$$= 14.7 \text{ m/s}$$

f) Using equation (3),

$$\text{distance fallen} = 14.7 \times 1$$
$$= 14.7 \text{ m}$$

g) The total distance fallen = 14.7 + 4.9 = 19.6 m

Example

Draw a velocity—time graph for a freely falling object for times up to 5 s. Use the graph to determine the distance fallen in 5 s.

Solution

The graph must have a constant slope of 9.8 m/s^2. This means the velocity must change by 9.8 m/s for every second. Figure 5.12 shows the graph. The distance fallen is the area under the velocity—time graph up to 5 s. This is half the area of the rectangle of sides about 50 m/s and 5 s, i.e.

$$\text{distance fallen} = \tfrac{1}{2} \times 50 \times 5 = 125 \text{ m}$$

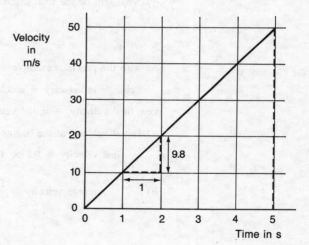

Fig. 5.12

Friction

When an object resting on the floor is given a push it may not move. If it does not move and accelerate then there can be no resultant force acting on the object. This means there must be another force exactly cancelling out the 'push' force. This opposing force is called the *frictional force*, see Fig. 5.13. Increasing the push force increases the frictional force so that they continue to cancel each other out. This continues until the object begins to slide. When this occurs the push force is greater than the frictional force and so there is a resultant force acting on the object.

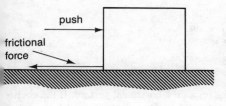

push

frictional
force

Fig. 5.13

Resultant force = push force − frictional force

There is a maximum value for the frictional force and once the push force is greater than this sliding occurs. This maximum frictional force is often called the limiting frictional force.

The size of the limiting frictional force depends on the size of the force acting at right angles to the surfaces in contact and responsible for keeping the surfaces in contact. This force is called the normal force. For an object on a horizontal plane the normal force is equal to the weight (in newtons) of the object. If the object is on an incline the normal force is equal to the component of the weight which is at right angles to the plane.

The size of the limiting frictional force does not depend on the areas of the surfaces in contact, but depends only on the normal force and the types of surface in contact.

Since

limiting frictional force ∝ normal force

then

limiting frictional force F = (a constant μ) × (normal force N)

i.e., $$F = \mu N$$

The constant, μ, is called the coefficient of friction. The term coefficient of static friction is used where the frictional force is the maximum value occurring when the object is just on the point of sliding.

When the object is in motion, frictional forces still exist and the same relationship holds, but the coefficient has a slightly smaller value. The coefficient is then called the dynamic coefficient of friction.

In summary:

1. Frictional forces are always in the opposite direction to motion or attempted motion.
2. The size of the limiting frictional force depends on the types of surfaces in contact.
3. The size of the limiting frictional force is independent of the areas of the surfaces in contact.
4. The size of the limiting frictional force is proportional to the normal force, $F = \mu N$.

Example

A block of steel of mass 5.0 kg rests on a horizontal surface. If the coefficient of static friction is 0.30 what horizontal force will need to act on the block to cause it to begin to slide?

Solution

Weight of block = mg = 5.0 × 9.8 = 49 N
Hence normal force = 49 N

Since $\qquad F = \mu N$
$$F = 0.30 \times 49$$
$$= 14.7 \text{ N}$$

Example

A block of wood of mass 2.0 kg rests on a horizontal surface. If the coefficient of static friction is 0.60 what will happen when horizontal forces of *a*) 5.0 N, *b*) 10.0 N, *C*) 15.0 N are applied to the block?

Solution

Weight of block $= mg = 2.0 \times 9.8 = 19.6$ N
Hence normal force $= 19.6$ N
limiting friction force $= \mu N = 0.60 \times 19.6$
$$= 11.8 \text{ N}$$

Hence the 5.0 N and the 10.0 N forces are insufficient to get the object sliding. Only the 15.0 N force will do this, since it is greater than the limiting frictional force.

Effects of friction

Many fastening devices rely on frictional forces to keep things fastened. For example, a screw relies on frictional forces between the screw material and the material in which it is screwed. Nuts, nails and clips also rely on frictional forces.

A car relies on frictional forces between the tyres and the road in order to move. Without such frictional forces the wheels would just rotate without moving the car along the road. Frictional forces are also necessary if the brakes are to stop a car.

The above are examples where frictional forces are useful, there are, however, many instances where such forces are not beneficial. Thus the frictional forces that occur in the bearings of rotating shafts means that energy has to be wasted which could otherwise more usefully be used in driving the shaft. Frictional forces also give rise to wear, for example, in the bearings of the shaft.

One way of reducing frictional effects is by lubrication. Effectively lubricants keep the rubbing surfaces apart and so avoid the frictional forces that occur between one solid surface sliding over another. There are, however, still resistance forces involved in the lubricant sliding over the solid surface, but they are considerably lower than those between the solid, unlubricated, surfaces.

Wave motion

Water waves are an obvious example of wave motion. Other examples are sound waves, light waves and radio waves. When a stone is dropped into a pond waves spread out across the surface of the water from the impact point. The wave motion does not involve the water moving outwards from the impact point. What happens is that the initial impact of the stone with the water causes the water surface at that point to be depressed, before it then springs back up again. This movement up and down affects the water surface immediately surrounding the impact point and it also moves up and down. This then affects the water further out and it then moves up and down. Hence the up-and-down motion of the water surface steadily moves outwards from the initial disturbance. We call the effect observed on the water surface a wave motion spreading out from the initial disturbance.

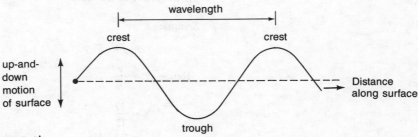

Fig. 5.14 Cross-section of a wave at some instant of time

Figure 5.14 shows the basic form of the water surface when a wave motion is occurring. The following are some basic terms used for waves.

1 *Wavelength* is the distance between two successive identical parts of a wave, e.g. between two crests. The symbol used for wavelength is λ (lambda) and the SI unit is the metre.

2 *Frequency* is the number of waves per second which will pass a point in the wave path, or it can be considered to be the number of waves produced per second. One wave is taken as being that part of the motion between two successive identical parts of the wave, e.g. between two crests. Instead of referring to the number of waves per second the term cycles per second is often used, a cycle being another term for a wave. The symbol used for frequency is f and the SI unit is the hertz (Hz). One hertz is one wave per second or one cycle per second.

3 The *velocity* of a wave is the velocity with which the wave motion moves outwards from initial disturbance, the velocity with which a crest or a trough appears to move. The symbol used is v and the SI unit is m/s.

If a disturbance results in three waves being produced every second then in one second the front of the wave motion will have travelled out a distance from the disturbance equal to three times the wavelength. If the wavelength is 20 mm then the disturbance travelled in 1 s is $3 \times 20 = 60$ mm. The velocity is the distance travelled per second and so the velocity in this case is 60 mm/s.

The above argument can be put in a more general form. If the frequency is f then f waves are produced per second. If each has a wavelength λ then the distance travelled in 1 s is $f\lambda$. Hence the velocity v is given by

$$v = f\lambda$$

Example

A water wave has a velocity of 10 m/s and a wavelength of 12 m. What is the frequency of the wave?

Solution

$$v = f\lambda$$

$$\text{Hence} \quad f = \frac{v}{\lambda}$$

$$= \frac{10}{12}$$

$$= 0.83 \text{ Hz}$$

Example

A radio wave is transmitted at a frequency of 1215 kHz and has a velocity of 300 000 000 m/s, (3.0×10^8 m/s). What is its wavelength?

Solution

$$v = f\lambda$$

$$\text{Hence} \quad \lambda = \frac{v}{f}$$

$$= \frac{300\ 000\ 000}{1215 \times 1000}$$

$$= 247 \text{ m}$$

Self-assessment questions

1 A person walks a distance of 8.0 km in 80 minutes. What was the person's average speed in m/s?
2 A train maintains a speed of 60 km/h for 20 minutes. How far will it have travelled in that time?
3 A cyclist travels 100 m in a northerly direction and then 100 m in a westerly direction. What will be the total distance covered and the displacement from the start point?
4 Explain the difference between scalar and vector quantities.
5 A car covers the first kilometre of a journey in 4 minutes and then takes a further 2 minutes for the next kilometre. What was its average speed over the entire 6 minutes?
6 A car travels at 60 km/h for 20 minutes and then 80 km/h for 10 minutes. What is the total distance travelled?
7 Draw the distance–time graph for the motion described by the following data:

Distance in m	0	10	20	30	40	50	60	70
Time in s	0	2	4	7	11	17	23	29

a) State whether the object was at rest, moving with constant speed or accelerating at the times of (*i*) 2 s, (*ii*) 5 s, (*iii*) 23 s.

b) Calculate the speed at (*i*) 2 s, (*ii*) 17 s.

8   Sketch the form of a distance—time graph for a car which starts from rest, accelerates and then moves with a constant speed before braking and coming to rest.

9   For the graph shown in Fig. 5.15, what is the average speed *a*) over the first 10 s, *b*) over the period 10 to 15 s, *c*) over the period 15 to 25 s, *d*) over the period 0 to 25 s?

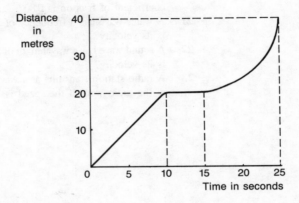

Fig. 5.15

10   A cyclist starting from rest reaches a velocity of 20 km/h after 12 s. What was the cyclist's average acceleration for that period?

11   A car accelerates from an initial velocity 50 km/h to 70 km/h in 2 s. What was its average acceleration?

12   A car travelling at 40 km/h is braked and comes to rest in 5.0 s. What was the average retardation?

13   The velocity of an object varies with time according to the following data.

Velocity in m/s	0	1.5	3.0	3.5	3.5	3.5
Time in s	0	1	2	3	4	5

a) Plot the velocity—time graph.

b) Is the object moving with constant velocity or accelerating at (*i*) 1 s, (*ii*) 4 s?

c) Estimate the acceleration at 1 s.

d) Estimate the distance covered by the object in 5 s.

14   The velocity of a car varies with time according to the following data.

Velocity in m/s	20	22	24	26	27	28	28
Time in s	0	1	2	3	4	5	6

a) Plot the velocity—time graph.

b) Is the car moving with constant velocity or accelerating at (*i*) 2 s, (*ii*) 5 s?

c) Estimate the acceleration at (*i*) 1 s, (*ii*) 3.5 s.

d) Estimate the distance covered by the car in (*i*) 1 s, (*ii*) 6 s.

15   A car starts from rest and accelerates uniformly at 1.0 m/s^2 for 10 s. It then moves with constant velocity for 50 s. Sketch the velocity—time graph for this motion.

16   A brick falls from a factory roof on to the ground 10 m below. Sketch the velocity—time graph for the fall and hence, or otherwise, determine the velocity attained by the brick on hitting the ground and the time taken for the fall.

17 A stone drops from the edge of a cliff. What will be its velocity after *a*) 2 s, *b*) 3 s?

18 A steel block of mass 6.0 kg rests on a horizontal surface. If the coefficient of static friction is 0.70 what horizontal force will be needed to start the block sliding?

19 If a force of 100 N is needed to start an object of mass 15 kg sliding along a horizontal chute, what is the coefficient of static friction?

20 What is the normal force that has to be applied between a brake pad and a steel disc if the frictional force required for braking is 1000 N and the coefficient of friction is 0.80?

21 A water wave has a frequency of 50 Hz and a wavelength of 3.0 mm, what is its velocity?

22 A sound wave has a wavelength of 200 mm and a frequency of 1.7 kHz, what is its velocity?

23 A radio station transmits at a wavelength of 200 m. What is the frequency of the radio wave if the speed is 300 000 000 m/s?

6 Energy

Work

Suppose you were to push a broken-down car along a road. You would be applying a force to the car and this would be causing the car to move through some distance. Work is said to be done if an object moves as a result of a force being applied to it. Pushing the car would certainly qualify as work! *Work* is defined as, see Fig. 6.1:

Work = force × distance moved in the direction of the force.

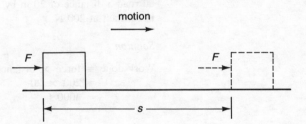

Fig. 6 .1 Work done = $F \times s$

When the unit of force is the newton and that of distance the metre, then the unit of work is the joule, (J).

$$1 \text{ J} = (1 \text{ N}) \times (1 \text{ m})$$

It is important to realise that the scientific definition of work given above is not quite the same as the every day use of that term. No work is done if a force is applied but the object does not move. So if you push the car but it does not move, no work is done. There has to be motion for work to occur. This is sometimes stated as: the point of application of the force must move through a distance for work to be done.

Example

Calculate the work done when a constant upward force of 100 N raises a load through a vertical height of 2.0 m.

73

Solution

Work done = force × distance moved in direction of force
= 100 × 2.0
= 200 J

Example

Calculate the work done when a hydraulic hoist steadily lifts a car of mass 1100 kg through a vertical height of 1.2 m.

Solution

Weight of car = 1100 × 9.8 = 10 780 N

The hoist must provide a force of 10 780 N vertically in order to overcome the weight of the car and lift it. Hence

Work done = force × distance moved in direction of force
= 10 780 × 1.2
= 12 936 J
= 12.936 kJ

Example

Calculate the work done in pulling a broken-down car steadily along a horizontal road a distance of 20 m by a horizontal rope if the tension in the rope is constant at 200 N.

Solution

Work done = force × distance moved in direction of force
= 200 × 20
= 4000 J

Work with an oblique force

Sometimes the motion of an object is not in the same direction as the applied force. To determine the work done by the force the distance moved in the direction of the force has to be determined, no other direction will do. Thus for the example shown in Fig. 6.2, the distance moved in the direction of the force is $s \cos \theta$. Hence

work = force × distance moved in direction of force
= $F \times s \cos \theta$

There is an alternative way of considering the situation and that is as

work = component of force in direction of
motion × distance moved

Thus for Fig. 6.3 the component of force in the direction of motion is $F \cos \theta$ (see Chapter 3 for discussion of components). Hence

work = $F \cos \theta \times s$

This is just the same as the answer arrived at by considering the distance moved in the direction of the force.

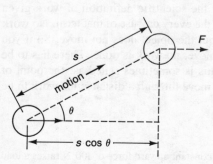

Fig. 6.2 Work done = $F \times s \cos \theta$

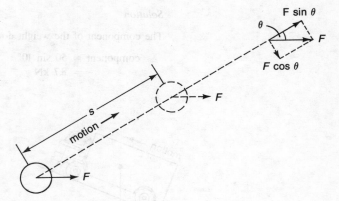

Fig. 6.3 Work done = $F \cos \theta \times s$

Example

A barge is pulled along a canal at a steady rate by a horizontal rope which
is at an angle of 30° to the direction of motion of the barge (Fig. 6.4). If the
tension in the rope is 500 N what is the work done in moving the barge 300 m?

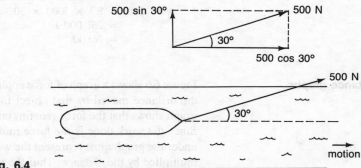

Fig. 6.4

Solution

Component of the force in the direction of motion = $F \cos 30°$
$$= 500 \times \cos 30°$$
$$= 433 \text{ N}$$

Hence   work done by the force = force component in direction of
                         motion × distance moved
$$= 433 \times 300$$
$$= 129\ 900 \text{ J}$$
$$= 129.9 \text{ kJ}$$

Example

A lorry is driven at a steady rate up a hill which is at 10° to the horizontal.
If the lorry has a weight of 50 kN what work is done in the lorry moving
30 m up the hill?

Solution

The component of the weight down the hill is, see Fig. 6.5:

component = 50 sin 10°
         = 8.7 kN

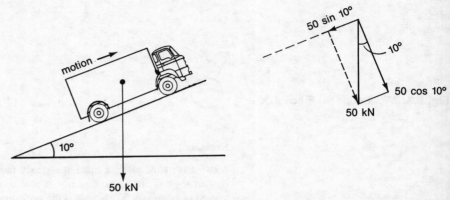

Fig. 6.5

This weight has to be overcome for the lorry to move steadily up the hill. Since

work done = component of force in direction of
            motion × distance moved
work done = 8.7 × 1000 × 30
          = 261 000 J
          = 261 kJ

Force−distance graphs

Figure 6.6 shows a graph of force applied to some object plotted against the distance moved by that object in the direction of the force. The graph shows that the force remains constant as the distance increases. Since the work done is the force multiplied by the distance, the area under the graph must represent the work done since it also is the force multiplied by the distance. Thus the work done when the force moves the object a distance of 4 m in the direction of the force is 10 × 4

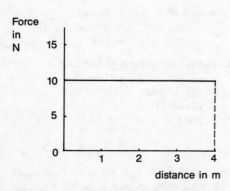

Fig. 6.6 A force-distance graph

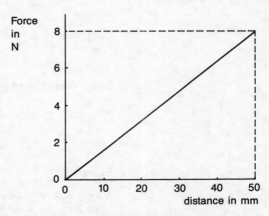

Fig. 6.7 A force-distance graph for a spring

= 40 J. Whatever the form of the force—distance graph the area under the graph is the work done by the force.

Figure 6.7 shows the type of graph that can be obtained by hanging different weights on the lower end of a vertically suspended spring and measuring the extension. The weight is the applied force and the extension the distance moved in the direction of the force. Thus the work done by a force in extending the spring by 50 mm is the area under the graph up to the 50 mm distance, i.e.

$$\text{area} = \tfrac{1}{2} \times \text{area of rectangle of sides 8 N and 50 mm}$$
$$= \tfrac{1}{2} \times 8 \times 50/1000$$
$$= 0.20 \text{ J}$$

Example

A car suspension requires a force of 500 N for each 10 mm it is compressed. Sketch a force—compression graph for the suspension up to a compression of 50 mm. Hence determine the work done in compressing the suspension by 50 mm.

Solution

Figure 6.8 shows the graph. Hence the area under the graph up to an extension of 50 mm is

$$\text{area} = \tfrac{1}{2} \times \text{area of rectangle of sides 2500 N and 50 mm}$$
$$= \tfrac{1}{2} \times 2500 \times 50/1000$$
$$= 62.5 \text{ J}$$

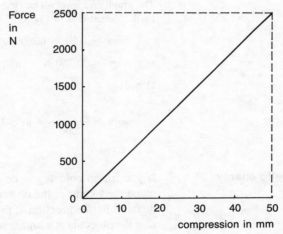

Fig. 6.8

Example

The following table gives the force acting on an object and the distance it has moved in the direction of the force at a number of instants. Plot a graph of force against distance and hence determine the work done by the force in moving the object a distance of 250 mm.

Force in N	20	40	60	50	40	30
Distance in mm	0	50	100	150	200	250

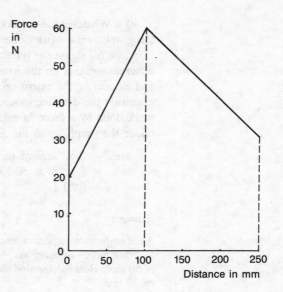

Fig. 6.9

Solution

Figure 6.9 shows the graph. The work done is the area under the graph up to a distance of 250 mm. This can be estimated by calculation or by counting the number of squares under the graph or alternatively calculating it using:

area of trapezium = $\frac{1}{2}$ (sum of parallel sides) $\times$ (distance of separation)

The graph area involves two trapeziums, indicated by the division of the area by the dotted line in Fig. 6.9. Thus.

area = $\frac{1}{2}(20 + 60)100 + \frac{1}{2}(60 + 30)150$

Area = 10 750 N $\times$ mm

Hence

work = $\dfrac{10\ 750}{1000}$ = 10.75 J

Potential and kinetic energy

If you lift an object, you do work in that you apply a force to overcome the weight of the object and move the object through a vertical distance in the direction of the applied force. At the end of the operation the object is at a higher level. The object is said to have acquired potential energy as a result of being lifted. If the object falls down from this height it does work since its weight is moving the object through a distance. Thus an object that is lifted up has the potential to be able to do work. *Potential energy* is the energy an object has by virtue of its position.

If you push an object along a horizontal plane and it moves it acquires a velocity. The object is said to have acquired kinetic energy as a result of the work being done on it giving it a velocity. When the pushing force on the object is stopped, it keeps moving for a short while and this is doing work against the frictional forces. *Kinetic energy* is the energy an object has by virtue of its motion.

An object is said to have *energy*, of whatever form, if it is capable of doing work.

Work as energy transfer

When you push an object and it moves, perhaps pushing a broken-down car along the road, you supply your energy and the end result is that the car gains energy. The work done is the energy transferred from you to the object. Whenever work is done there is a transfer of energy from one object to another.

Work done = energy transferred

Doing work is like drawing money out of a bank. The money is transferred from the bank to your pocket. The work can be considered as the process of drawing out the money, the money representing energy. In drawing the money from the bank, the money leaving the bank is equal to the money gained by your pocket. Money is conserved in the transaction. The same situation occurs with energy. When work is done energy is transferred from one object to another with the energy lost by one object being equal to the energy gained by the other. Energy is conserved.

The *law of conservation of energy* states that in any energy transfer the total amount of energy remains constant.

Forms of energy

The law of conservation of energy means that whenever a force does work something has lost energy and something else has gained energy. Thus when a battery is connected to an electric motor and causes the motor to lift some object, the battery must be transferring energy to the motor which then does work and results in a gain in potential energy for the lifted object. The battery is said to have chemical energy. This is transformed to electrical energy to drive the motor and hence to potential energy.

The battery could have been used to light a lamp. Then the chemical energy is used to give thermal energy and radiant energy, i.e. the lamp filament gets hot and glows.

There are many forms of energy, e.g. chemical energy (energy released as a result of a chemical reaction), mechanical energy, (kinetic and potential energy), thermal energy, electrical energy, nuclear energy, and radiant energy (the energy associated with electromagnetic waves such as light and radio).

Power

Power is the rate at which energy is transferred from one object to another. Thus if work is the means of energy transfer, then the power is the rate of doing work. The unit of power is the watt (W) and this is a transfer of 1 joule per second.

The average power over a period of time is

$$\text{average power} = \frac{\text{work done}}{\text{time}}$$

A further, useful, equation can be developed from this. Since

$$\text{work done} = \text{force} \times \text{distance moved in direction of force}$$

$$\text{average power} = \frac{\text{force} \times \text{distance moved in direction of force}}{\text{time}}$$

But the velocity in the direction of the force is

$$\text{velocity} = \frac{\text{distance moved in direction of force}}{\text{time}}$$

Thus average power = force × velocity in direction of force

Example

Calculate the power required to steadily lift a car of mass 1000 kg on a hydraulic hoist through a vertical height of 2.0 m in 30 s.

Solution

Force required to overcome weight = 1000 × 9.8 N

work done = force × distance moved in direction of force

= 1000 × 9.8 × 2.0 J

Hence average power = $\dfrac{1000 \times 9.8 \times 2.0}{30}$

= 653 W

Example

A builder's hoist steadily lifts a mass of 100 kg through a vertical height of 25 m. What power will be required if the lift is to take 50 s?

Solution

Force required to overcome weight = 100 × 9.8 N

work done = force × distance moved in direction of force

= 100 × 9.8 × 25 J

Hence average power = $\dfrac{100 \times 9.8 \times 25}{50}$

= 490 W

Example

A car travelling along a level road at a steady speed of 60 km/h suffers a total resistance to motion of 1.5 kN. What is the power required by the car to maintain this steady speed?

Solution

$$60 \text{ km/h} = \frac{60 \times 1000}{60 \times 60} \text{ m/s}$$

$$\text{average power} = \text{force} \times \text{velocity}$$

$$= (1.5 \times 1000) \times \left(\frac{60 \times 1000}{60 \times 60} \right)$$

$$= 25\,000 \text{ W} = 25 \text{ kW}$$

Example

What power will be required by a car to pull a trailer at 80 km/h along a level road if the force in the tow bar is 500 N?

Solution

$$80 \text{ km/h} = \frac{80 \times 1000}{60 \times 60} \text{ m/s}$$

$$\text{average power} = \text{force} \times \text{velocity}$$

$$= 500 \times \left(\frac{80 \times 1000}{60 \times 60} \right)$$

$$= 11\,111 \text{ W} = 11.1 \text{ kW}$$

Efficiency

Efficiency is defined as

$$\text{efficiency} = \frac{\text{useful output energy}}{\text{input energy}}$$

The symbol used for efficiency is η (eta). Efficiency has no units since it is a ratio of quantities with the same units. It is often expressed as a percentage.

$$\text{Efficiency} = \frac{\text{useful output energy}}{\text{input energy}} \times 100\%$$

A machine with an efficiency of 60% would transform 60% of its input energy into useful output energy, the remaining 40% being wasted in overcoming friction and other losses. A perfect machine would have an efficiency of 100%.

Example

A machine supplied with an energy of 4000 J is able to lift a mass of 50 kg through a vertical height of 5.0 m. What is the efficiency of the machine?

Solution

The force required to overcome the weight $= 50 \times 9.8$ N

$$\text{work done} = \text{force} \times \text{distance moved in direction of force}$$
$$= 50 \times 9.8 \times 5.0 \text{ J}$$

This work done is the useful output energy, hence

$$\text{efficiency} = \frac{\text{useful output energy}}{\text{input energy}} \times 100\%$$

$$= \frac{50 \times 9.8 \times 5.0}{4000} \times 100\%$$

$$= 61\%$$

Example

A motor has an input power of 2.0 kW and gives a useful power output of 1.6 kW. What is its efficiency?

Solution

The energy transferred per second is the power, thus the input energy in 1 s is 1.6 kJ and the useful output energy in 1 s is 2.0 kJ. Hence

$$\text{efficiency} = \frac{\text{useful output energy}}{\text{input energy}} \times 100\%$$

$$= \frac{1.6}{2.0} \times 100\%$$

$$= 80\%$$

Heat and temperature

Heat is the transfer of energy between two points that occurs because the two points are at different temperatures. The unit of heat is the joule (J).

Heat and work are the means by which energy transfers can occur between two objects. The result of such transfers are changes in the energies of the object receiving and giving the transfer. These changes may be a change in chemical energy, potential energy, kinetic energy, electrical energy, thermal energy, etc.

The transfer of energy to an object can affect the temperature of that body. Temperature is the degree of hotness of an object. Heat and temperature are not the same thing. In some instances the energy transfer into an object might not result in any temperature change of the object, e.g. when it causes ice at 0 °C to change to water at 0 °C. Temperature is measured by instruments called thermometers.

Temperatures are expressed as numbers on a scale, the most commonly used scales being the Celsius and Kelvin scales of temperature. The *Celsius scale* has the melting point of ice as 0 °C and the boiling point of water as 100 °C. Temperatures on the *Kelvin scale* have the same size degree as the Celsius scale, so

a change of 1 °C = a change of 1 K

The Kelvin scale has its zero value so defined that the melting point of ice is 273.15 K and the boiling point of water 373.15 K. Thus

temperatures on Kelvin scale = temperatures on Celsius scale + 273.15

Example

What are the following Celsius temperatures on the Kelvin scale?

   a) − 10 °C, *b*) 10 °C, *c*) 120 °C.

Solution

To convert these to temperatures on the Kelvin scale, add 273.15. Hence

   a) − 10 °C = − 10 + 273.15 = 263.15 K
   b)   10 °C = 10 + 273.15 = 283.15 K
   c)   120 °C = 120 + 273.15 = 393.15 K

Heat capacity

Heat capacity is the quantity of heat required to raise the temperature of a body by 1 °C (or 1 K). The symbol C is used for heat capacity and it has the unit J/°C or J/K.

$$\text{Heat capacity } C = \frac{Q}{t}$$

where Q is the quantity of heat and t the rise in temperature. Thus if an object has a heat capacity of 2000 J/K then to produce a rise in temperature of 10°C there must be a heat input of

$$\begin{aligned}
Q &= C \times t \\
&= 2000 \times 10 \\
&= 20\ 000 \text{ J} \\
&= 20 \text{ kJ}
\end{aligned}$$

The *specific heat capacity* is the heat required to raise the temperature of 1 kg of the substance by 1 °C (or 1 K). The symbol c is used for specific heat capacity and it has the unit J/(kg °C) or J/(kg K). This unit is also written as J kg^{-1} °C^{-1} or J kg^{-1} K^{-1}.

$$\text{Specific heat capacity } c = \frac{Q}{m \times t}$$

where Q is the heat required to raise the temperature of m kg of the substance by t °C (or t K).

The following are some typical values of specific heat capacities.

Substance	c in J kg^{-1} K^{-1}
water	4200
ice	2100
aluminium	950
iron	500
copper	390

Example

Calculate the heat required to raise the temperature of an object by 12 °C if it has a heat capacity of 400 J/K.

Solution

$$\text{Heat capacity } C = \frac{Q}{t}$$

$$\begin{aligned}
\text{Hence} \quad Q &= C \times t \\
&= 400 \times 12 \\
&= 4800 \text{ J}
\end{aligned}$$

Example

Calculate the heat required to raise the temperature of 2.0 kg of copper by 30 °C if it has a specific heat capacity of 390 J kg^{-1} K^{-1}.

Solution

$$\text{Specific heat capacity } c = \frac{Q}{m \times t}$$

$$\begin{aligned}
\text{Hence} \quad Q &= m \times c \times t \\
&= 2.0 \times 390 \times 30 \\
&= 23\ 400 \text{ J}
\end{aligned}$$

Example

Calculate the temperature change produced when 5000 J of heat are supplied to 1.5 kg of aluminium. The specific heat capacity of aluminium is 950 J kg^{-1} K^{-1}.

Solution

$$\text{Specific heat capacity } c = \frac{Q}{m \times t}$$

$$\begin{aligned}
\text{Hence} \quad t &= \frac{Q}{m \times c} \\
&= \frac{5000}{1.5 \times 950} \\
&= 3.5 \text{ K (or °C)}
\end{aligned}$$

Example

500 g of water are in a container of heat capacity 50 J/K. What will be the rise in temperature of the water after 600 J of heat are supplied? Specific heat capacity of water = 4200 J kg^{-1} K^{-1}.

Solution

The heat will increase the temperature of both the water and the container. Part of the 600 J will heat the water and the remainder the container.

For the water:

specific heat capacity $c = \dfrac{Q_1}{m \times t}$

and so $\qquad\qquad\qquad Q_1 = m \times c \times t$

and so
For the container:

Heat capacity $C = \dfrac{Q_2}{t}$

and so $\qquad\qquad Q_2 = C \times t$

and so

Since heat supplied $= 600 = Q_1 + Q_2$

then $\qquad\qquad 600 = \left(\dfrac{500}{1000} \times 4200 \times t\right) + (50 \times t)$

$600 = 2100\,t + 50\,t$

$t = \dfrac{600}{2150}$

$= 0.28°C$

Change of state

Figure 6.10 shows how the temperature of two objects might change with time when there is a steady rate of heat input to the two objects. In the one case, Fig. 6.10(a), the temperature rises at a steady rate as time progresses. This object might be a block of metal with the temperature changes being from about 20 °C to 80 °C. In the other case, Fig. 6.10(b), the temperature does not rise uniformly with time. Indeed there is a period of time when, though there is heat entering the object, there is no change in temperature at all. This is an object which

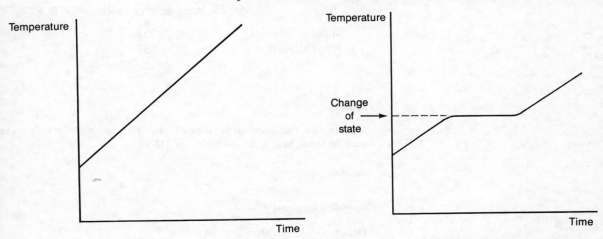

a) When no change of state occurs

b) With a change of state

Fig. 6.10 Temperature–time graphs

is changing state, perhaps paraffin wax which melts at about 50 °C.

The term *change of state* is used when a substance changes from solid to liquid, or liquid to gas. Solid, liquid and gas are called states of matter. To convert a solid to a liquid, or a liquid to a gas, energy has to be supplied. This energy is used for changing the state and not for changing the temperature. For this reason it is called *latent heat*, the word latent meaning 'hidden'. Heat which results in a temperature change is called *sensible heat*.

When a gas turns into a liquid, or a liquid into a solid, the change of state requires the latent heat to be removed from the substance for the change to occur.

The *specific latent heat L* is the energy required to change the state of 1 kg of a substance.

$$\text{Specific latent heat } L = \frac{Q}{m}$$

where Q is the energy needed to change the state of m kg of the substance without any temperature change occurring.

The term specific latent heat of fusion is used when the change of state is from solid to liquid. The term specific latent heat of vaporisation is used when the change of state is from liquid to gas. The following are some typical specific latent heat values.

Substance	Specific latent heat of fusion in kJ/kg
Water	335
Aluminium	387
Copper	205
Iron	268

	Specific latent heat of vaporisation in kJ/kg
Water	2257
Ethyl alcohol	857

Example

Calculate the heat required to change 10 kg of ice into water at 0 °C. The specific latent heat of fusion for ice is 335 kJ/kg

Solution

$$\text{Specific latent heat } L = \frac{Q}{m}$$

Hence
$$\begin{aligned} Q &= L \times m \\ &= 335 \times 1000 \times 10 \\ &= 3350\ 000 \text{ J} \\ &= 3.35 \text{ MJ} \end{aligned}$$

Example

Calculate the heat required to change 200 g of water at 100 °C into steam. The specific latent heat of vaporisation for water is 2257 kJ/kg.

Solution

Specific latent heat $L = \dfrac{Q}{m}$

Hence $\qquad Q = L \times m$

$\qquad\qquad\qquad = 2257 \times 1000 \times 200/1000$

$\qquad\qquad\qquad = 451\ 400\ \text{J}$

$\qquad\qquad\qquad = 451.4\ \text{kJ}$

Example

Calculate the heat required to change 500 g of water at 20 °C into steam at 100 °C. Specific heat capacity of water = 4200 J kg^{-1} K^{-1}, specific latent heat of vaporisation of water = 2257 kJ/kg.

Solution

This problem involves two steps, one the calculation of the heat required to raise the temperature of the water from 20 °C to 100 °C without any change of state, and secondly the heat required to change the water at 100 °C to steam at the same temperature. For the first step:

specific heat capacity $c = \dfrac{Q}{m \times t}$

Hence $\qquad Q = m \times c \times t$

$\qquad\qquad\qquad = \dfrac{500}{1000} \times 4200 \times (100 - 20)$

$\qquad\qquad\qquad = 168\ 000\ \text{J}$

$\qquad\qquad\qquad = 169\ \text{kJ}$

For the second step:

specific latent heat $L = \dfrac{Q}{m}$

Hence $\qquad Q = L \times m$

$\qquad\qquad\qquad = 2257 \times 1000 \times \dfrac{500}{1000}$

$\qquad\qquad\qquad = 1128\ 500\ \text{J}$

$\qquad\qquad\qquad = 1128\ \text{kJ}$

Hence the total heat required is 1128 + 168 = 1296 kJ.

Expansion of solids

Solids generally expand when their temperature increases. The amount of expansion depends on the amount by which the temperature changes. It also depends, for a given temperature change, on the original length of the solid. The *linear expansivity*, or coefficient of linear expansion, of a material is defined as

$$\text{linear expansivity} = \frac{\text{change in length}}{\text{original length} \times \text{change in temperature}}$$

The unit of linear expansivity is /°C or /K (or K^{-1}).

The following are some typical values of linear expansivities for solids.

Substance	Linear expansivity in /K
Aluminium	0.000 023
Copper	0.000 017
Mild Steel	0.000 011
Brass	0.000 019
Brick	0.000 005
Soda glass	0.000 009

Example

A bar of copper of length 200 mm at 0 °C is heated to 50 °C. If the linear expansivity is 0.000 017 /K calculate the amount by which the bar expands.

Solution

$$\text{Linear expansivity} = \frac{\text{change in length}}{\text{original length} \times \text{change in temperature}}$$

Hence   change in length = linear expansivity × original length × change in temperature
= 0.000 017 × 200 × 50
= 0.17 mm

Example

A telephone cable is fixed between two posts 40 m apart. If the temperature when the cable is fixed between the posts was 25 °C how much slack should the engineers allow if the cable is not to become taut, and possibly break, for temperatures as low as − 10 °C? Linear expansivity of the cable is 0.000 017 /K.

Solution

The slack is the amount by which the cable will contract when the temperature drops from 25 °C to − 10 °C, i.e. a temperature change of 35 °C. Hence, as

$$\text{linear expansivity} = \frac{\text{change in length}}{\text{original length} \times \text{change in temperature}}$$

change in length = linear expansivity × original length × change in temperature

At −10 °C the length is 40 m, hence, when the temperature rises by 35 °C,

$$\text{change in length} = 0.000\ 017 \times 40 \times 35$$
$$= 0.0238 \text{ m}$$
$$= 23.8 \text{ mm}$$

Volume expansivity

When a solid expands, not only does its length change but so also does its volume. The *cubic expansivity*, or coefficient of cubic expansion, is defined as

$$\text{cubic expansivity} = \frac{\text{change in volume}}{\text{original volume} \times \text{change in temperature}}$$

The unit of volume expansivity is /°C or /K (or K^{-1}). To a reasonable approximation the volume expansivity for a solid is three times its linear expansivity.

Hollow bodies expand as though they were completely solid throughout. Thus a glass beaker would expand as though it were a completely solid block of glass.

Liquids generally expand when there is an increase in temperature and cubic expansivity is defined in the same way for them as for solids. The term real cubic expansivity is used for a liquid when referring to the actual volume and change in volume of the liquid. When the liquid is in a container, both the container and the liquid will expand when heated and the apparent volume of the liquid indicated by the liquid level is an underestimate of the true volume. We thus use the term apparent cubic expansion for a liquid in a container of a particular material.

$$\text{real cubic expansivity} = \text{apparent cubic expansivity} +$$
$$\text{cubic expansivity of the}$$
$$\text{container material}$$

Gases can expand when there is an increase in temperature and we can use the term cubic expansivity to describe their volume change. This is provided the temperature change results only in a volume change and there is no pressure change occurring.

Example

A block of metal has a volume of 0.05 m^3 at 20 °C. What will be its change in volume at 100 °C if the linear expansivity of the metal is 0.000 020 /K?

Solution

$$\text{Cubic expansivity} = \frac{\text{change in volume}}{\text{original volume} \times \text{change in temperature}}$$

Hence change in volume = cubic expansivity × original volume ×
change in temperature
$$= 3 \times 0.000\ 020 \times 0.05 \times (100 - 20)$$
$$= 0.000\ 24 \text{ m}^3$$

Example

A glass container can hold 1000 cm^3 of a liquid at 20 °C. What volume will it be able to hold at 60 °C if the linear expansivity of the glass is 0.000 009 /K?

Solution

$$\text{Cubic expansivity} = \frac{\text{change in volume}}{\text{original volume} \times \text{change in temperature}}$$

Hence   change in volume = cubic expansivity × original volume ×
change in temperature

= 3 × 0.000 009 × 1000 × (60 − 20)

= 1.08 cm^3

Hence   volume = 1000 + 1.08

= 1001.08 cm^3

Example

If the cubic expansivity of water (at about room temperature) is 0.000 21/K what will be the change in volume of 1000 cm^3 of water when the temperature rises from 20 °C to 30 °C?

Solution

$$\text{Cubic expansivity} = \frac{\text{change in volume}}{\text{original volume} \times \text{change in temperature}}$$

Hence   change in volume = cubic expansivity × original volume ×
change in temperature

= 0.000 21 × 1000 × (30 − 20)

= 2.1 cm^3

Example

A glass container will hold 1000 cm^3 of liquid at 20 °C. How much water should be put into the container at 20 °C if the container is to be full at 80 °C (at higher temperatures it overflows)? The cubic expansivity of water can be taken as 0.000 21 /K and the linear expansivity of the glass 0.000 009 /K.

Solution

Real cubic expansivity = apparent cubic expansivity +
cubic expansivity of container

Hence   apparent cubic expansivity = real cubic expansivity −
cubic expansivity of container

= 0.000 21 − (3 × 0.000 009)

= 0.000 183/K

This value can be used to calculate the change in volume allowing the container changes with temperature to be ignored. At 80 °C the apparent volume is 1000 cm^3 thus it is necessary to calculate the volume when the liquid is cooled to 20 °C.

$$\text{Apparent cubic expansivity} = \frac{\text{apparent change in volume}}{\text{original volume} \times \text{change in temperature}}$$

Hence   apparent change in volume = apparent cubic expansivity ×
original volume × change in
temperature
= 0.000 183 × 1000 × (80 − 20)
= 10.98 cm³

Hence, volume of water to be put into the container at 20 °C is

volume = 1000 − 10.98
= 989.02 cm³

Practical applications and implications of expansion

The following are some of the practical applications and implications
of materials expanding when their temperature increases and contract-
ing when it decreases.

1. Telephone cables and overhead electrical transmission lines are
   suspended between supports so that they are slack in summer.
   Otherwise when winter comes and they contract they could become
   taut and break.
2. The length of a bridge depends on the temperature and to allow
   for expansion and contraction the ends of large bridges are often
   supported on rollers to allow them to expand and contract freely.
3. Riveting makes use of expansion and contraction. A hot, expanded
   rivet is inserted through two holes in the metal plates being joined.
   The ends of the hot rivet are then hammered into a mushroom
   shape. As the rivet cools it contracts and pulls the two plates
   together.
4. Fitting a metal collar onto a metal shaft is often achieved by heating
   the metal collar so that it expands and is bigger than the shaft.
   Then it is slid over the shaft and when it cools and contracts, a
   tight fit is obtained. The alternative to heating the collar is to cool
   the shaft.
5. An item which finds many applications is the bimetallic strip. This
   consists of two strips of different metals welded or riveted together,
   see Fig. 6.11. The different metals expand by different amounts
   for the same rise in temperature. But they are fixed together. The
   result is that the compound strip curves, the material with the
   higher linear expansivity, and hence greater amount of expansion,
   being on the outside of the curve and the lower linear expansivity

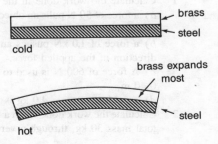

Fig. 6.11   The bimetallic strip

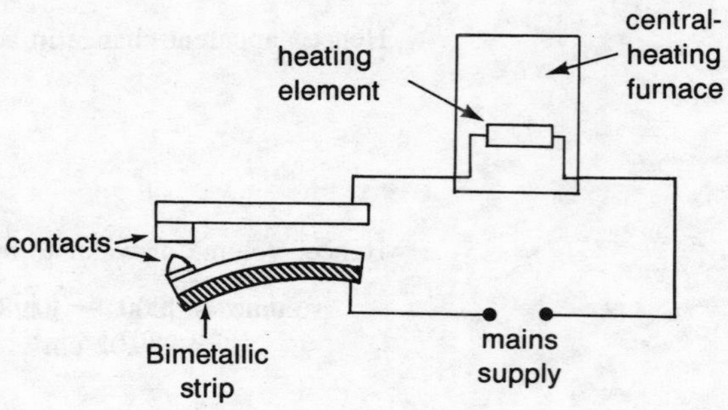

Fig. 6.12 A bimetallic strip thermostat

one on the inside. Such bimetallic strips are frequently used in thermostats, switching off an electric current when the required temperature is reached, see Fig. 6.12.

6. The mercury-in-glass thermometer relies on the expansion of mercury, the height of a column of mercury up a scale being a measure of temperature.

7. When water is cooled it contracts until a temperature of 4 °C is reached. Then further cooling results in the water expanding, until at 0°C the water begins to freeze. On freezing there is a large increase in volume. A consequence of this is that if water freezes in a pipe the forces exerted on the pipe by the expanding water can cause the pipe to burst.

8. When a liquid or a gas increases in temperature it expands and so the volume occupied by a given mass of liquid or gas is significantly increased. This results in a decrease in density (density = mass/volume). Thus when, for example, water is heated in a cooking pan on the cooker, the water at the bottom of the pan increases in temperature and becomes less dense than the water higher up in the pan. The result is that the lower-density, hotter water rises up through the higher-density, colder water and there is movement of water in the pan. This movement is called a convection current. Such convection currents can be used in domestic hot-water systems to circulate hot water from the top of the hot-water cylinder and allow cold water to enter at the bottom. Such convection currents also occur in the Earth's atmosphere and give rise to winds.

Self-assessment questions

1 Calculate the work done in the following situations:
 a) a force of 50 N pulls an object through a distance of 1.2 m in the direction of the force,
 b) a force of 1.0 kN pushes an object through a distance of 50 cm in the direction of the applied force,
 c) A force of 600 N is used to vertically lift an object through a height of 0.40 m,
 d) an object of mass 4.0 kg is lifted vertically through 2.0 m.

2 Calculate the work done when a builders' hoist is used to lift a load of bricks, total mass 30 kg, through a vertical height of 60 m.

3 Calculate the work done in the following situations:
   a) a force of 100 N causes an object to move through a distance of 30 cm in a direction at an angle of 30° to the direction of the force,
   b) a force of 1.4 kN causes an object to move through a distance of 1.1 m in a direction at an angle of 45° to the direction of the force.

4 What is the work done when a car of mass 1000 kg is steadily driven 30 m up an incline which is at 10° to the horizontal?

5 Sketch the force—distance graphs for the following situations and in each case determine the work done in producing a displacement of 50 mm.
   a) A constant force of 50 N acts over the distance of 50 mm.
   b) A constant force of 20 N acts as the distance increases from 0 to 20 mm and then the force abruptly increases to 50 N for the remaining distance up to 50 mm.

c) Force in N	0	10	20	30	40	50
Distance in mm	0	10	20	30	40	50
d) Force in N	10	15	20	30	40	50
Distance in mm	0	10	20	30	40	50

6 Explain what is meant by *a*) potential energy and *b*) kinetic energy.

7 A machine lifts a weight of 500 N through a vertical height of 0.60 m in 2.0 s. Calculate *a*) the work done and *b*) the power required.

8 Calculate the power required for a pumping engine to raise 50 kg of water per minute from a well 4 m deep.

9 A conveyor belt is used to lift 20 castings, each of mass 40 kg, through a height of 4.0 m in 2.0 minutes. Calculate the power required.

10 What is the total resistance to motion experienced by a car if a power of 20 kW is required to maintain a steady speed of 50 km/h?

11 A motor has an input power of 2.0 kW. If it has an efficiency of 60% what will be *a*) the output power, *b*) the work done in 2.0 minutes?

12 A hoist lifts a weight of 500 N through a height of 6.0 m in 10 s. What is *a*) the work done in the time, *b*) the output power of the hoist, *c*) the input power if the hoist has an efficiency of 70%?

13 Calculate the heat required to raise the temperature of a casting by 200 °C if it has a heat capacity of 2500 J/K.

14 Calculate the heat required to raise the temperature of 500 g of water from 20 °C to 100 °C. The specific heat capacity of water is 4200 J kg^{-1} K^{-1}.

15 Calculate the heat lost by a block of iron of mass 12 kg in cooling from 300 °C to 20 °C if the specific heat capacity of iron is 500 J kg^{-1} K^{-1}.

16 Distinguish between sensible heat and latent heat.

17 Calculate the energy required to change 12 kg of ice at 0 °C to water at 0 °C. Specific latent heat of fusion of water is 335 kJ/kg.

18 Calculate the heat required to convert 2.0 kg of water at 20 °C into steam at 100 °C. Specific heat capacity of water = 4200 J kg^{-1} K^{-1}, specific latent heat of vaporisation of water is 2257 kJ/kg.

19 Calculate the heat required to melt a block of iron of mass 500 kg initially at 20 °C. Melting point of iron = 1200 °C, specific heat capacity of iron = 0.50 kJ kg^{-1} K^{-1}, specific latent heat of fusion of iron = 270 kJ/kg.

20 Calculate the linear expansion of two bridges of span 100 m when the temperature rises by 20°C if one bridge is made of concrete and the other of steel. Linear expansivity of concrete = 0.000 014/K, linear expansivity of steel = 0.000 011/K.

21 If a cable is 100 m long at 20 °C how long will it be at 25 °C? Linear expansivity of cable = 0.000 017 /K.

22 A domestic central-heating system uses a continuous length of copper piping of length 7.0 m. By how much will the pipe expand when the water in the pipe raises the pipe temperature from 15 °C to 70 °C? Linear expansivity of copper = 0.000 017 /K.

23 A flask is completely filled with 1000 cm^3 of water at 15 °C. How much of the water will overflow when the flask is heated to 70 °C? Apparent cubic expansivity of the water = 0.000 18 /K.

24 A steel tank with a volume of 4.0 m^3 at 15 °C is filled with paraffin at that temperature. How much of the paraffin will overflow when the temperature rises to 25 °C? Linear expansivity of steel = 0.000 011 /K, cubic expansivity of paraffin = 0.00120/K.

25 Explain how a bimetallic strip thermostat operates.

7 Electricity

Basic terms

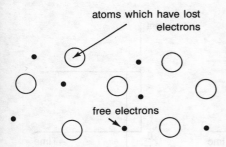

Fig. 7.1 A model of a metal

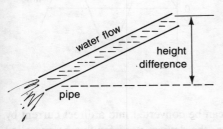

Fig. 7.2 A model for electric current

The following are basic terms used in discussions of electricity.

1 The *electron* is a constituent of all atoms and has a negative charge.

2 *Metals* are materials with atoms which readily lose electrons. At room temperature most of the atoms in a metal have lost an electron and these free electrons exist within the metal without attachment to any particular atom, see Fig. 7.1.

3 A *conductor* is a material through which charge can be readily moved. Metals are good conductors because the electrons, which are negatively charged, can easily be made to move in a particular direction through the material.

4 An *insulator* is a material which has virtually no charge which can be moved through the material. It has atoms which have their electrons tightly bound to them.

5 An *electric current* occurs when charge moves in a particular direction through a material.

6 A *voltage* or *potential difference* is needed to make a current occur in a material, i.e. to make charge flow in a particular direction. A conductor can be considered to be like a water pipe with the water being the charge. To obtain movement of the water through the pipe it can be tilted so that there is a height difference between the two pipe ends, see Fig. 7.2. This height difference is the equivalent of the potential difference needed to cause charge movement in a conductor.

7 The unit of current is the *ampere* (A) or amp for short, and the unit of voltage or potential difference, is the *volt* (V).

8 Current is measured using instruments called *ammeters* and voltage is measured using *voltmeters*.

9 *Direct current*, d.c., is when the current through a conductor is maintained in the same direction. For the electric current model described in Fig. 7.2, the water only flows in one direction.

10 *Alternating current*, a.c., is when the current through a conductor is alternating between flowing first in one direction and then in the opposite direction. Considering electric current in terms of water flow through a pipe, the height difference between the ends of the pipe is

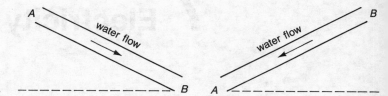

Fig. 7.3 A model for a.c.

continually being changed so that first end *A* is higher than end *B*, then end *B* is higher than end *A*, see Fig. 7.3. To produce the alternating water flow there is an alternating height difference. To produce an alternating current, an alternating potential difference, i.e. voltage, is required.

11   A *battery* is a source of voltage which gives rise to a direct current, see Fig. 7.4(*a*).

12   The *electric mains supply* in Britain is a source of voltage which gives rise to alternating current, see Fig. 7.4(*b*).

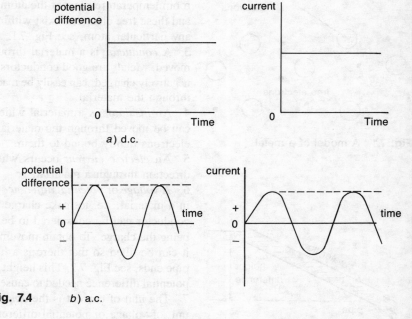

Fig. 7.4   *a*) d.c.

b) a.c.

13   An alternating current can be converted into a direct current by a process called *rectification*, the devices used for this being called *rectifiers*. For applications such as electric fires or electric lamps it does not matter which way the current flows through the item and so a.c. can be used. There are, however, some items where the current must only flow in one direction, e.g. many electronic systems, and for these either a battery must be used or the a.c. mains supply must be passed through a rectifier.

14   The *direction of current* in a circuit is taken to be from the positive terminal of a battery or power supply to the negative terminal.

Current

Current is the rate of flow of charge past a point in a conductor. The unit of charge is the coulomb (C) and the current is one ampere (1A) when one coulomb passes a point in one second. The average current over a period of time is given by:

$$\text{average current} = \frac{\text{charge passing point}}{\text{time}}$$

Thus when 2 C of charge pass a point in 50 s then the average current over that time is 2/50 = 0.04 A.

Example

What is the current in a conductor when 10 C pass through the conductor every minute?

Solution

$$\text{Average current} = \frac{\text{charge passing point}}{\text{time}}$$

The charge has to be in coulombs and the time in seconds for the current to be in amperes. Thus

$$\text{average current} = \frac{10}{60}$$
$$= 0.17 \text{ A}$$

Example

When there is a current of 150 mA in a conductor how much charge is passing a point in that conductor in 30 s?

Solution

$$150 \text{ mA} = \frac{150}{1000} \text{ A}$$

$$\text{average current} = \frac{\text{charge passing point}}{\text{time}}$$

Hence charge passing point = average current × time

$$= \frac{150}{1000} \times 30$$
$$= 4.5 \text{ C}$$

Electric circuit symbols

Figure 7.5 shows some of the common symbols used in circuit diagrams (according to British Standard 3939 and general international usage). A line is used to represent a conductor. Where two conductors cross without electrical connection the two lines can cross. If, however, there is an electrical connection when the conductors cross then, unless

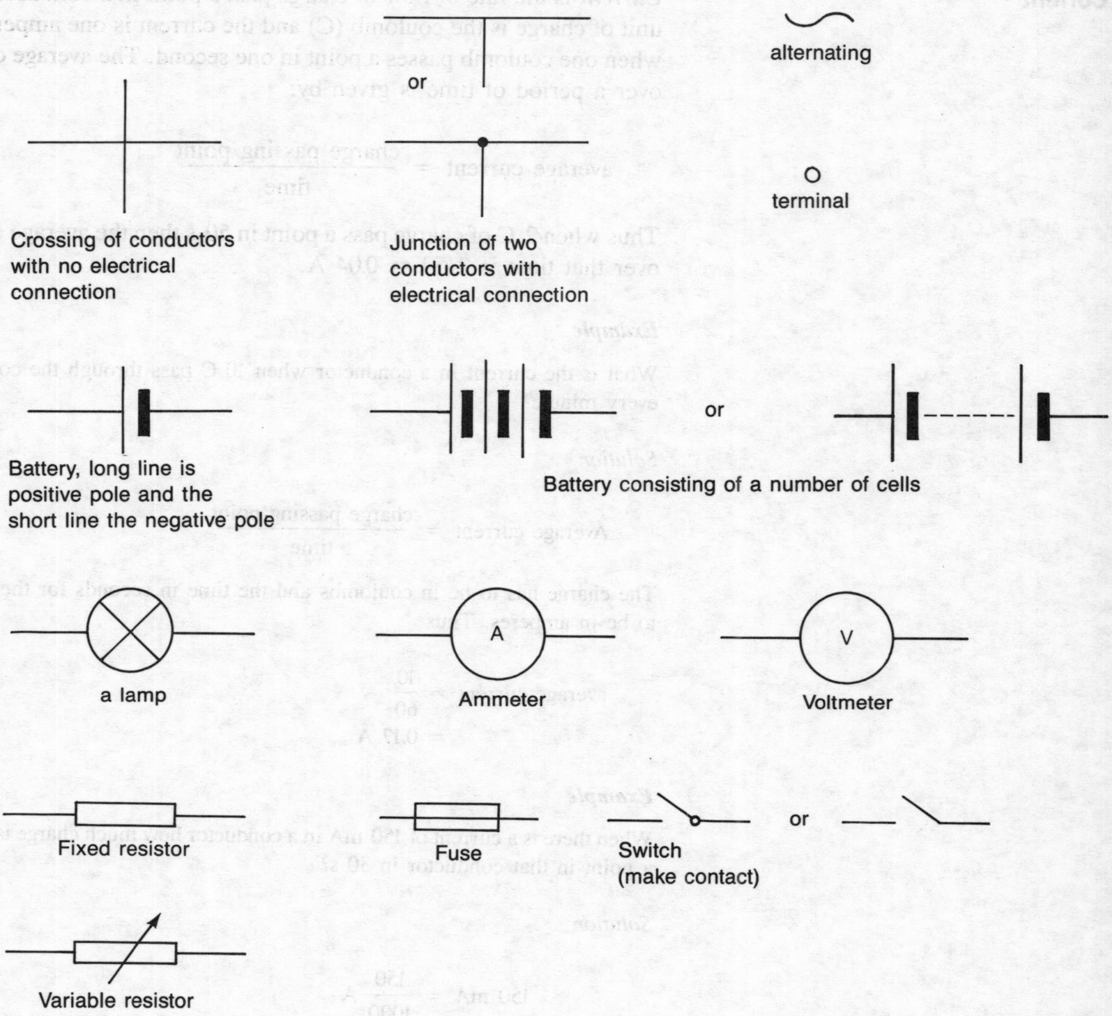

alternating

or

terminal

Crossing of conductors
with no electrical
connection

Junction of two
conductors with
electrical connection

Battery, long line is
positive pole and the
short line the negative pole

or

Battery consisting of a number of cells

a lamp

Ammeter

Voltmeter

Fixed resistor

Fuse

Switch
(make contact)

or

Variable resistor

Fig. 7.5 Circuit symbols

there is an obvious junction, a dot can be placed at that point. Lines
from symbols should be continued for a short distance before they
meet another symbol or a connecting line.

Resistance

Figure 7.6 shows an electrical circuit by which the current through
a circuit element and the potential difference across it can be measured.
The ammeter is connected in series with the circuit element so that
the same current flows through it as through the circuit element. The
voltmeter is connected in parallel with the circuit element. In terms
of the water-flow model used for current in Fig. 7.2, the voltmeter

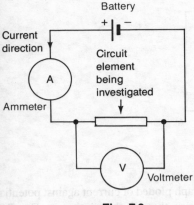

Current
direction

Battery

Circuit
element
being
investigated

A

Ammeter

V   Voltmeter

Fig. 7.6

is measuring the height difference between the two ends of the element.

The potential difference across the element divided by the current through it is called the *resistance*.

$$\text{Resistance } R = \frac{\text{potential difference } V}{\text{current } I}$$

$$R = \frac{V}{I}$$

The unit of resistance is the ohm (Ω) when the potential difference is in volts and the current in amperes. The bigger the resistance of a circuit element the bigger the potential difference needed to drive a given current through it. A circuit element which is made to offer a certain resistance is called a resistor.

Example

What is the resistance of a resistor which passes a current of 120 mA when the potential difference across it is 2.0 V?

Solution

$$R = \frac{V}{I}$$

$$= \frac{2.0}{(120/1000)}$$

$$= 16.7\ \Omega$$

Example

A resistor of resistance 10 Ω is connected in a circuit so that the potential difference across it is 20 V. What is the current through it?

Solution

$$R = \frac{V}{I}$$

$$\text{Hence } I = \frac{V}{R}$$

$$= \frac{20}{10}$$

$$= 2.0\ \text{A}$$

Ohm's law — an investigation

Figure 7.7 shows a circuit that could be used to determine the relationship between the current through a circuit element and the potential difference across it. The current through the element is varied by means of the variable resistor and readings are taken of the current and potential difference.

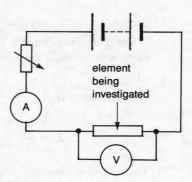

Fig. 7.7 Investigating potential difference-current relationships

For some circuit elements a graph plotted of current against potential difference is a straight line passing through the origin, see Fig. 7.8. Such a circuit element is said to be a linear device and obey Ohm's law.

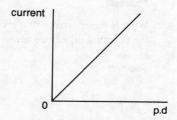

Fig. 7.8 An element obeying Ohm's law

Ohm's law can be stated as — provided physical conditions, such as temperature, do not change, then over a wide range of potential differences the current is proportional to the potential difference, i.e.,

current ∝ potential difference

This can be written as

potential difference = a constant × current

This constant is called the resistance. Hence

potential difference = resistance × current

Resistance is a constant if the element obeys Ohm's law, not changing when the potential difference changes.

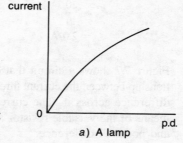
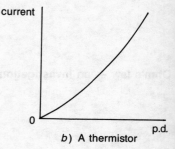

Fig. 7.9 Examples of elements not obeying Ohm's law

a) A lamp

b) A thermistor

Ohm's law is obeyed by many materials, particularly metals. However, there are devices which do not obey Ohm's law. Figure 7.9 shows two examples.

Example

If the current through a resistor is 120 mA when the potential difference across it is 6.0 V, what will be the current when the potential difference is 3.0 V if the resistor obeys Ohm's law?

Solution

Since current ∝ potential difference

halving the potential difference must mean the current is halved, i.e. to 60 mA.

Example

Figure 7.10 shows on the same graph axes the current potential difference relationships for two materials.

a) Which of the materials obeys Ohm's law?
b) What is the resistance of the material that obeys Ohm's law?
c) Sketch a graph showing how the resistance of the material that does not obey Ohm's law varies with potential difference.

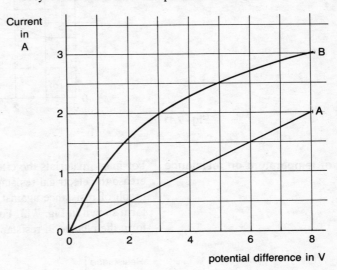

Fig. 7.10

Solution

a) Material A obeys Ohm's law since the graph is a straight line passing through the origin.

b) $R = \dfrac{V}{I}$

The current at any potential difference value can be taken and the resistance calculated.

Hence since when $R = 8V$, $I = 2A$, (values taken from the graph):

$$R = \frac{8}{2} = 4\ \Omega$$

c) The resistance of material B changes as the potential difference changes. Thus at p.d. = 1 V the current is 1 A and so $R = V/I = 1/1 = 1\ \Omega$
At p.d. = 2 V the current is 1.6 A and so $R = V/I = 2/1.6 = 1.25\ \Omega$
At p.d. = 4 V the current is 2.3 A and so $R = V/I = 4/2.3 = 1.7\ \Omega$
At p.d. = 6 V the current is 2.7 A and so $R = V/I = 6/2.7 = 2.2\ \Omega$
At p.d. = 8 V the current is 3.0 A and so $R = V/I = 8/3.0 = 2.7\ \Omega$

Figure 7.11 shows the graph of resistance against potential difference.

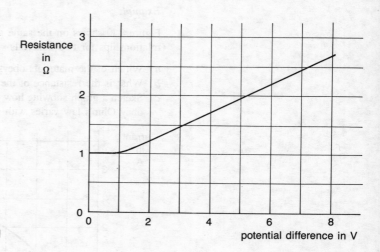

Fig. 7.11

Effect of temperature on resistance

For many materials the effect of an increase in temperature is to increase the electrical resistance. This is often the case with metals. A graph of resistance against temperature for a typical metal is of the form shown in Fig. 7.12. For materials showing a relationship of that form the change in resistance depends on the change in temperature

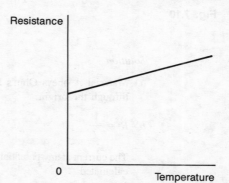

Fig. 7.12 Effect of temperature on resistance for a typical model

and the size of the initial resistance. We can define a *temperature coefficient of resistance* as

$$\frac{\text{change in resistance}}{\text{original resistance} \times \text{change in temp.}}$$

The unit of temperature coefficient of resistance is /°C or /K (or K^{-1}]. The following are some typical values at about room temperature.

Substance	Temp. coefficient of resistance in /K
Aluminium	0.0038
Copper	0.0043
Constantan	almost zero
Manganin	0.00001

Example

Using the values of the temperature coefficients of resistance given in the above table calculate:

a) the resistance of a coil of copper wire at 25 °C if it has a resistance of 5.0 Ω at 20 °C,

b) the resistance of a coil of manganin wire at 25 °C if it has a resistance of 100 Ω at 20 °C,

c) the resistance of a coil of constantan wire at 25 °C if it has a resistance of 20 Ω at 20 °C.

Solution

Temp. coefficient of resistance = $\dfrac{\text{change in resistance}}{\text{original resistance} \times \text{change in temp.}}$

Hence change in resistance = temp. coefficient of resistance × original resistance × change in temp.

a)   change in resistance = 0.0043 × 5.0 × (25 − 20)

         = 0.11 Ω

Hence resistance = 5.11 Ω

b)   change in resistance = 0.00001 × 100 × (25 − 20)

         = 0.005 Ω

Hence resistance = 100.005 Ω

c) Since the temperature coefficient of resistance is zero then the resistance does not change with temperature and so the resistance is still 20 Ω

Series and parallel circuits

In a *series circuit* the same current flows through each of the circuit elements. Thus if we have a series connections of lamps, as shown in Fig. 7.13, the same current must pass through each lamp regardless of whether each has the same resistance. If, however, one of the lamps burns out and ends up with a broken filament all the lamps go out since there is no path for the current through the lamps.

In a *parallel circuit* the same potential difference exists across each element. Thus for a parallel connection of lamps, as shown in

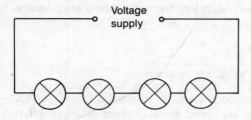

Fig. 7.13 Series connection of lamps

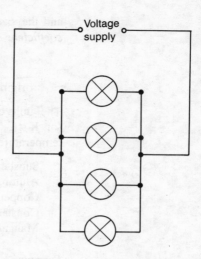

Fig. 7.14 Parallel connection of lamps

Fig. 7.14, there is the same potential difference across each lamp regardless of whether each has the same resistance. If one of the lamps burns out and ends up with a broken filament only that one lamp goes out; all the others remain on. This is because there are still current paths through these lamps.

Christmas tree lights are generally connected in series. Thus when one of the lamps breaks all go out. The lights in a house are connected in parallel so that each can be independently switched on or off.

Series and parallel connection of resistors

Consider three resistors connected in series, see Fig. 7.15. The same current will pass through each resistor. The potential difference across all the resistors V is equal to the sum of individual potential differences across each resistor. Thus

$$V = V_1 + V_2 + V_3$$

But $V_1 = IR_1$, $V_2 = IR_2$, and $V_3 = IR_3$. Thus

$$V = IR_1 + IR_2 + IR_3$$

$$\frac{V}{I} = R_1 + R_2 + R_3$$

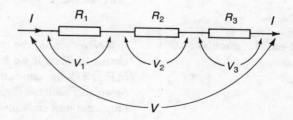

Fig. 7.15 Resistors in series

If the three resistors are replaced by a single resistor R giving the same current I when potential difference V is applied, then $V/I = R$. Thus the combined resistance of the three resistors when connected in series is:

$$R = R_1 + R_2 + R_3$$

Consider three resistors connected in parallel, see Fig. 7.16. The same potential difference is across each resistor. The current I entering the parallel arrangement must equal the sum of the currents through each resistor. Thus

$$I = I_1 + I_2 + I_3$$

But $I_1 = V/R_1$, $I_2 = V/R_2$ and $I_3 = V/R_3$. Hence

$$I = \frac{V}{R_1} + \frac{V}{R_2} + \frac{V}{R_3}$$

$$\frac{I}{V} = \frac{1}{R_1} + \frac{1}{R_2} + \frac{1}{R_3}$$

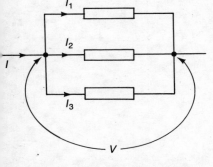

Fig. 7.16 Resistors in parallel

If the three resistors are replaced by a single resistor giving the same current I when potential difference V is applied then since $V/I = R$, $I/V = 1/R$. Thus the combined resistance of the three resistors connected in parallel is:

$$\frac{1}{R} = \frac{1}{R_1} + \frac{1}{R_2} + \frac{1}{R_3}$$

Example

What is the total resistance for the following arrangements of resistors?
a) Resistors of 10 Ω, 20 Ω and 30 Ω connected in series.
b) Resistors of 10 Ω and 20 Ω connected in parallel.

Solution

a) For resistors connected in series

$$R = R_1 + R_2 + R_3$$
$$\text{Hence} \quad R = 10 + 20 + 30$$
$$= 60 \ \Omega$$

b) For two resistors connected in parallel

$$\frac{1}{R} = \frac{1}{R_1} + \frac{1}{R_2}$$

$$\text{Hence} \quad \frac{1}{R} = \frac{1}{10} + \frac{1}{20}$$

$$\frac{1}{R} = \frac{2 + 1}{20}$$

$$R = \frac{20}{3} = 6.7 \ \Omega$$

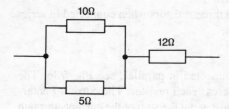

Fig. 7.17

Example

Calculate the total resistance of the arrangement of resistors shown in Fig. 7.17.

Solution

Consider first that part of the circuit with the two resistors connected in parallel. For two resistors connected in parallel

$$\frac{1}{R} = \frac{1}{R_1} + \frac{1}{R_2}$$

Hence
$$\frac{1}{R} = \frac{1}{10} + \frac{1}{5}$$

$$\frac{1}{R} = \frac{1 + 2}{10}$$

$$R = \frac{10}{3} = 3.3 \ \Omega$$

The circuit is now equivalent to a 3.3 Ω resistor conected in series with a 12 Ω resistor. For two resistors connected in series

$$R = R_1 + R_2$$
$$= 3.3 + 12$$
$$= 15.3 \ \Omega$$

Electromotive force (e.m.f.)

A device is said to be a source of *electromotive force*, usually abbreviated to e.m.f., if it is capable of making currents occur in circuits. It does this by maintaining a potential difference between its terminals. Thus a battery has an e.m.f. because it produces a potential difference between its terminals which can drive a current through a circuit. Similarly an electric generator or dynamo is a source of e.m.f. The unit of e.m.f. is the volt.

The *e.m.f.* is equal to the potential difference between the terminals of a source when no current is taken from it. When a current is taken from the source, because it is connected to an external circuit, then the potential difference between the terminals drops. This is because some of the energy is used to drive a current through the source itself.

p.d. between terminals = e.m.f. − p.d. used to drive current through the source itself

The potential difference used to drive the current through the source is zero when the current taken is zero and increases as the current taken from the source increases, see Fig. 7.18.

The source is considered to have an internal resistance and thus when there is a current taken from it there is a potential difference produced across this internal resistance. The greater the current taken the greater the potential difference across the internal resistance and so the less the potential difference available between the terminals for the external circuit.

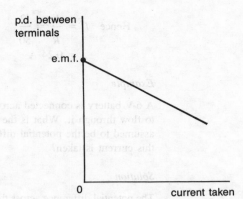

p.d. between terminals

e.m.f.

0   current taken

Fig. 7.18 Variation of the potential difference between the terminals of a source with the current taken

The p.d. used to drive current through source itself $= Ir$ where I is the current being taken and r the internal resistance. Thus

p.d. between terminals = e.m.f. $- Ir$

The greater the internal resistance of a source the less the potential difference obtainable between its terminals for an external circuit.

Circuit calculations

The following examples are based on the use of the following equations:

$$\text{Resistance } R = \frac{\text{potential difference } V}{\text{current } I} \tag{1}$$

or rearranging this gives

$$V = IR \tag{2}$$

For resistors of resistances R_1, R_2 and R_3 connected in series

$$\text{total equivalent resistance } R = R_1 + R_2 + R_3 \tag{3}$$

If the resistors are connected in parallel, the total equivalent resistance R is given by

$$\frac{1}{R} = \frac{1}{R_1} + \frac{1}{R_2} + \frac{1}{R_3} \tag{4}$$

Example

A 12 V battery has a 50 Ω resistor connected between its terminals. What will be the current through the resistor? Assume that the 12 volts is the potential difference between the battery terminals when this current is taken.

Solution

The potential difference across the resistor is 12 V, thus using equation (1) gives:

$$R = \frac{V}{I}$$

Hence $I = \dfrac{V}{R} = \dfrac{12}{50}$

$= 0.24$ A

Example

A 6-V battery is connected across a resistor and causes a current of 10 mA to flow through it. What is the resistance of the resistor if the 6 V can be assumed to be the potential difference between the battery terminals when this current is taken?

Solution

The potential difference across the resistor is 6 V, thus using equation (1) gives

$$R = \frac{V}{I}$$

$$= \frac{6}{(10/1000)}$$

$$= 600 \ \Omega$$

Example

For the circuit shown in Fig. 7.19, determine a) the current in the circuit and b) the potential difference across the 10 Ω resistor.

Solution

a) For resistors connected in series, using equation (3) gives

$$R = R_1 + R_2 + R_3$$
$$= 2 + 5 + 10$$
$$= 17 \ \Omega$$

From equation (1),

since $R = \dfrac{V}{I}$

then $I = \dfrac{V}{R} = \dfrac{2}{17}$

$= 0.12$ A

b) Using equation (2), applied just to the 10 Ω resistor:

$V = IR$

$= 0.12 \times 10$

$= 1.2$ V

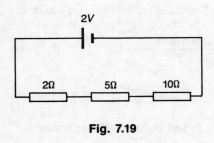

Fig. 7.19

Example

For the circuit shown in Fig. 7.20 determine a) the total circuit resistance and b) the current I.

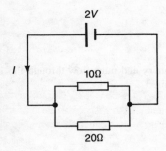

Fig. 7.20

Solution

a) Using equation (4) for resistors connected in parallel

$$\frac{1}{R} = \frac{1}{R_1} + \frac{1}{R_2}$$

$$\frac{1}{R} = \frac{1}{10} + \frac{1}{20}$$

$$\frac{1}{R} = \frac{2 + 1}{20}$$

$$R = 6.7 \ \Omega$$

b) Using equation (1)

$$R = \frac{V}{I}$$

Hence $\quad I = \dfrac{V}{R} = \dfrac{2}{6.7}$

$$= 0.30 \ A$$

Example

For the circuit shown in Fig. 7.21 calculate *a*) the total circuit resistance, *b*) the current through the 10 Ω resistor, *c*) the potential difference across the 10 Ω resistor, and *d*) the current through the 6 Ω resistor.

Solution

a) For the resistors connected in parallel, using equation (4)

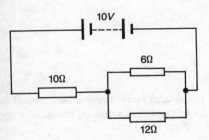

Fig. 7.21

$$\frac{1}{R} = \frac{1}{R_1} + \frac{1}{R_2}$$

$$\frac{1}{R} = \frac{1}{6} + \frac{1}{12}$$

$$\frac{1}{R} = \frac{2 + 1}{12}$$

$$R = 4 \ \Omega$$

This resistance can then be added to the 10 Ω resistance, since they are connected in series. Using equation (3)

$$R = R_1 + R_2$$
$$= 10 + 4$$
$$= 14 \ \Omega$$

b) The current in the circuit of equivalent resistance 14 Ω is given by equation (1):

$$R = \frac{V}{I}$$

Hence $I = \dfrac{V}{R} = \dfrac{10}{14}$

$\qquad = 0.71$ A

This is the current drawn from the battery and must flow through the 10 Ω resistor and hence is the answer required.

c) Using equation (2)

$V = IR$
$\quad = 0.71 \times 10$
$\quad = 7.1$ V

d) The potential difference across the parallel arrangement of 6 Ω and 12 Ω must be

$\qquad$ p.d. across 6 Ω = total p.d. − p.d. across 10 Ω

This is because the total p.d of 10 V is applied across both the 10 Ω resistor and the parallel arrangement.

$\qquad$ Hence p.d. across 6 Ω = 10 − 7.1
$\qquad\qquad\qquad\qquad\qquad\quad = 2.9$ V

Using equation (1), the current through the 6 Ω resistor is given by:

$$R = \frac{V}{I}$$

$$I = \frac{V}{R} = \frac{2.9}{6}$$

$$\quad = 0.48 \text{ A}$$

Incidentally, the current entering the parallel arrangement must be equal to the sum of the currents through the 6 Ω and 12 Ω resistors, i.e. the current through the 12 Ω resistor is 0.71 − 0.48 = 0.23 A.

Example

For the circuit shown in Fig. 7.22 calculate the value of the resistance *R* if the current is 120 mA.

Solution

For resistors in series, using equation (3):

$\qquad$ total resistance = 5 + *R*

From equation (1)

$\qquad$ total resistance = $\dfrac{V}{I}$

$$5 + R = \frac{12}{(120/1000)} = 100$$

$$R = 100 - 5$$
$$\quad = 95 \ \Omega$$

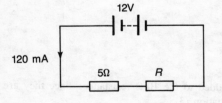

Fig. 7.22

120 mA

12V

5Ω   R

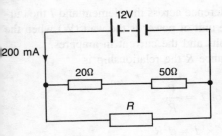

Fig. 7.23

Example

For the circuit shown in Fig. 7.23 calculate the value of the resistance R and the current through it.

Solution

Using equation (3), the combined resistance for the two resistors in series is,

resistance = 20 + 50 = 70 Ω

For this resistance in parallel with R, using equation (4), gives:

$$\frac{1}{\text{total resistance}} = \frac{1}{70} + \frac{1}{R}$$

$$= \frac{R + 70}{70 \times R}$$

Hence   total resistance $= \dfrac{70 \times R}{R + 70}$

The potential difference applied across this resistance is 12 V and the current through it is 200 mA, and from equation (1)

$$\text{total resistance} = \frac{V}{I}$$

Hence, $\dfrac{70 \times R}{R + 70} = \dfrac{12}{(200/1000)} = 60$

$$70 \times R = (60 \times R) + (60 \times 70)$$
$$(70 \times R) - (60 \times R) = 60 \times 70$$
$$10 \times R = 60 \times 70$$
$$R = 420 \ \Omega$$

The potential difference of 12 V is applied across this resistance, using equation (1)

since $R = \dfrac{V}{I}$

then $I = \dfrac{V}{R} = \dfrac{12}{420}$

$$= 0.029 \text{ A}$$

Power

In an electrical circuit energy is transferred from the source of the e.m.f., perhaps a battery, to the elements connected in the electrical circuit by a current. A resistor increases its temperature when the current passes through it. A motor in the circuit could do mechanical work. Whatever the circuit element the rate at which energy is transferred to it, i.e. the *power*, is given by

power = VI

where V is the potential difference across the element and I the current passing through it. The unit of power is the watt (W) when the potential difference is in volts and the current in amperes.

For a resistor with resistance R the relationship is

$$R = \frac{V}{I} \quad \text{or} \quad V = IR \quad \text{or} \quad I = \frac{V}{R}$$

$$\text{Thus} \quad \text{power} = VI = (IR) \times I = I^2R$$

$$\text{or} \quad \text{power} = VI = V \times \left(\frac{V}{R}\right) = \frac{V^2}{R}$$

Example

Calculate the power consumed when a current of 2.0 A passes through a filament lamp of resistance 10 Ω.

Solution

$$\begin{aligned}\text{Power} &= I^2R \\ &= 2.0^2 \times 10 \\ &= 40 \text{ W}\end{aligned}$$

Example

Calculate the power consumed when a current of 150 mA passes through a resistor and produces a potential difference of 4.0 V across it.

Solution

$$\text{Power} = VI$$

$$= 4.0 \times \frac{150}{1000}$$

$$= 0.60 \text{ W}$$

Example

A 12 V battery has a 100 Ω resistor connected across it. Calculate the power consumed by the resistor.

Solution

$$\text{Power} = \frac{V^2}{R}$$

$$= \frac{12^2}{100}$$

$$= 1.4 \text{ W}$$

Example

Calculate the current taken by a 60 W lamp when connected to a 240 V supply.

Solution

$$\text{Power} = VI$$

$$\text{Hence } I = \frac{\text{power}}{V}$$

$$= \frac{60}{240}$$

$$= 0.25 \text{ A}$$

Fuses

Fuses are devices designed to break an electrical connection at some particular value of current. Thus, for example, a fuse specified as a 13 A fuse is designed to break when the current reaches 13 A. This stops the current rising to a value greater than the 13 A and so damaging either the elements in the circuit or the circuit cables themselves.

Fuses are made of materials which melt when the specified current is reached. On melting they break the electrical connection.

Example

The following fuses are available: 2 A, 5 A, 10 A and 13 A. Which would be the most appropriate fuse to fit when the following appliances are connected to a 240 V supply?
a) A table lamp rated at 60 W.
b) An electric fire rated at 3 kW.
c) A toaster rated at 1 kW.

Solution

$$\text{Power} = VI$$

$$\text{Hence } I = \frac{\text{power}}{V}$$

a) $I = \dfrac{60}{240}$

$$= 0.25 \text{ A}$$

The 2 A fuse would be the most appropriate.

b) $I = \dfrac{3000}{240}$

$$= 12.5 \text{ A}$$

The 13 A fuse would be the most appropriate.

c) $I = \dfrac{1000}{240}$

$$= 4.2 \text{ A}$$

The 5 A fuse would be the most appropriate.

Self-assessment questions

1   What are the SI units of *a*) current, *b*) potential difference, *c*) power, *d*) resistance?

2   What is the difference between d.c. and a.c.?

3   What charge passes a point in a circuit per second when the current is 2 A?

4   A current of 4 A flows in a circuit for 5 minutes. What charge has been moved in that time?

5   What is the resistance of a resistor which passes a current of 0.50 A when the potential difference across it is 10 V?

6   A resistor of resistance 20 Ω has a current of 120 mA passing through it. What is the potential difference across it?

7   What is the current through a 12 Ω resistor when it has a potential difference of 6.0 V across it?

8   *a*) State Ohm's law.

    b) What form will a graph be of current through an element against potential difference across it if it obeys Ohm's law?

    c) For an element that obeys Ohm's law what will happen to the potential difference across it if the current through it is doubled?

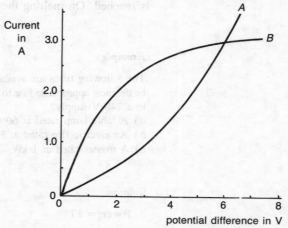

Fig. 7.24

9   Figure 7.24 shows two graphs, on the same axes of current against potential difference, for different circuit elements *A* and *B*. What are the resistances of *A* and *B* at currents of *a*) 1.0 A, *b*) 3.0 A?

10 Figure 7.25 shows two graphs, on the same axes, showing how the resistance of two different circuit elements varies with potential difference. *a*) Which of the elements obeys Ohm's law? *b*) Sketch the current–potential difference graph for element *A*.

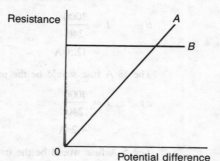

Fig. 7.25

11 Identify whether the following are connected in series or parallel.
   a) The circuit has ten lamps and when one lamp breaks all the lamps go out.
   b) The circuit has six lamps and when one lamp breaks the other lamps remain on.

12 A 5 Ω resistor, a 10 Ω resistor and a 15 Ω resistor are connected in series.
   a) Will the current be the same through each resistor?
   b) Will the potential difference across each resistor be the same?

13 A 10 Ω resistor and a 20 Ω resistor are connected in parallel.
   a) Will the current be the same through each resistor?
   b) Will the potential difference across each resistor be the same?

14 What is the total resistance in each of the following arrangements of resistors?
   a) 10 Ω in series with 20 Ω.
   b) 10 Ω in parallel with 20 Ω.
   c) 2 Ω, 4 Ω and 6 Ω all connected in series.

15 Calculate the total resistance in each of the arrangements of resistors shown in Fig. 7.26.

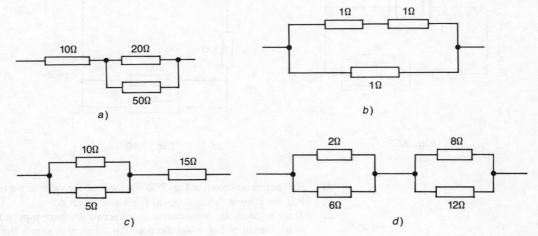

Fig. 7.26

a)   *b*)   *c*)   *d*)

16 A 12 V battery is connected across a resistor of resistance 10 Ω. What will be *a*) the current through the resistor and *b*) the potential difference across it if the 12 V can be regarded as the potential difference between the battery terminals?

17 A 10 Ω and a 15 Ω resistor are connected in series across a 12 V supply. What will be *a*) the current in the circuit and *b*) the potential difference across the 15 Ω resistor?

18 For the circuit shown in Fig. 7.27 determine the currents I, I_1 and I_2.

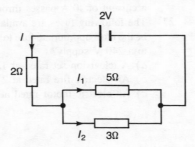

Fig. 7.27

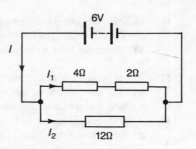

Fig. 7.28

19 For the circuit shown in Fig. 7.28 determine the currents I, I_1 and I_2 and the potential difference across the 4 Ω resistor.

20 For the circuit shown in Fig. 7.29 calculate the value of the resistance R if the current I is to be 2.0 A.

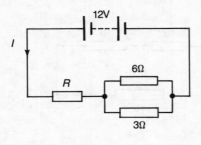

Fig. 7.29

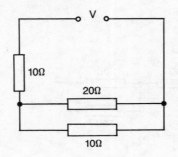

Fig. 7.30

21 For the circuit shown in Fig. 7.30 what will be the value of the supply voltage V if the current through the 10 Ω resistor is 1.5 A?

22 Three resistors are connected in series across a voltage supply. If the current in the circuit is 2 A when the potential differences across the resistors are 3 V, 5 V and 8 V, what are a) the supply voltage, b) the total circuit resistance, c) the resistances of the three resistors?

23 A 100 W lamp is connected to a 240 V supply. Calculate a) the current in the lamp, b) its resistance and c) the energy consumed by the lamp in one hour.

24 Calculate the power consumed when a current of 4 A passes through a resistor of resistance 5 Ω.

25 A 6 V battery has a 50 Ω resistor connected across it. Calculate the power consumed by the resistor.

26 Calculate the power consumed by an electric fire of resistance 25 Ω when a current of 10 A passes through it.

27 The following fuses are available: 2 A, 5 A, 10 A and 13 A. Which would be the most appropriate fuse to fit when the following appliances are connected to a 240 V supply?

a) A television set rated at 120 W.

b) An electric fire rated at 2 kW.

c) An electric motor rated at 500 W.

8 Effects of electricity

Heating effect

When a current passes through a conductor its temperature increases. The electric fire is an obvious example of this. The rate at which electrical energy is converted into heat is the power, with

$$\text{power} = VI = I^2R = \frac{V^2}{R}$$

where I is the current through the conductor, V the potential difference across it and R its resistance. The unit of power is the watt. (See previous chapter for more details.)

The total amount of energy converted in a time t when the power is P during that time interval is

$$\text{energy} = P \times t$$

The energy will be in joules when the power is in watts and the time in seconds.

Another unit of energy frequently used in electricity supply is the kilowatt hour. This is the unit used for electricity bills. One *kilowatt hour* is the amount of energy used in 1 hour when the power is 1 kW. Thus if you use an electric fire rated at 3 kW for 5 hours the number of units, i.e. kilowatt hours, used is

$$
\begin{aligned}
\text{energy used} &= \text{power} \times \text{time} \\
&= 3 \times 5 \\
&= 15 \text{ kilowatt hours.}
\end{aligned}
$$

In joules, since 1 kW = 1000 W and 1 hour = 60 × 60 s,

$$
\begin{aligned}
1 \text{ kilowatt hour} &= 1000 \times 60 \times 60 \\
&= 3\ 600\ 000 \text{ J}
\end{aligned}
$$

Example

In a household during a day a 3 kW fire is used for 5 hours, a washing machine of 1 kW is used for 1 hour and 6 lights, each at 100 W, are used for an average of 5 hours each. How many kilowatt hours of electricity are used in the day?

Solution

For the fire, energy used = 3 × 5 = 15 kilowatt hours
For the washing machine, energy used = 1 × 1 = 1 kilowatt hour

For the electric lamps, energy used = $6 \times \dfrac{100}{1000} \times 5 = 3$ kilowatt hours

Total energy used = 15 + 1 + 3 = 19 kilowatt hours

Chemical effect

For a material to be a good conductor of electricity it must have particles free to move through it and carry charge. In the case of metals, which are good conductors, the free charge carriers are electrons, see Chapter 7. Insulators have virtually no free charge carriers.

Figure 8.1 shows an arrangement by which the conduction of electric currents by liquids can be investigated. When distilled water is used there is virtually no current, the distilled water acting as an insulator (distilled water is very pure water). However, if common salt, sodium chloride, is dissolved in the water it becomes a good conductor of electricity.

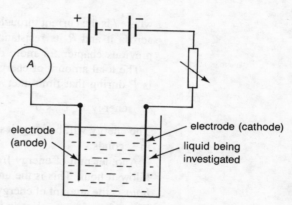

electrode (anode)

electrode (cathode)

liquid being investigated

Fig. 8.1 Investigating conduction by liquids

Distilled water contains virtually no particles which are free to carry charge, hence no current is possible. Dissolving sodium chloride in water introduces particles which are free and carry charge. The sodium chloride is considered to have broken apart into two parts, one carrying a positive charge and the other a negative charge. The positive part is a sodium atom with one electron missing, hence it has a net positive charge. The chlorine part is a chlorine atom with one extra electron, the electron the sodium atom lost. The result is that the chlorine atom has a net negative charge. These atoms with extra electrons or short of electrons are called *ions*.

The strips or rods of good conducting materials dipping into the liquid, as shown in Fig. 8.1, are called *electrodes* and are the means by which a potential difference is connected across the liquid. The electrode connected to the positive side of the battery, or d.c. supply, is called the *anode* while the electrode connected to the negative side is called the *cathode*. With a potential difference applied across the liquid, the positive ions move towards the negative electrode, i.e. the cathode, and the negative ions move towards the positive electrode, i.e. the anode. This movement of ions constitutes the current. A liquid which conducts electricity by the movement of ions is called an *electrolyte*.

When a current is passed through copper sulphate solution with copper electrodes, copper is deposited on the cathode and lost from the anode. The copper sulphate on dissolving in water breaks up into copper and sulphate ions. The copper ions are positively charged and the sulphate ions negatively charged. When a potential difference is connected across the solution the positive copper ions move to the cathode and become deposited on it. The negative sulphate ions move to the anode and combine with the copper of the electrode to produce copper sulphate. This effect has a number of applications. It is used for copper plating objects, i.e. coating them with a thin layer of copper. It is used in refining copper in that if the cathode is initially a thin strip of pure copper and the anode a block of impure copper the copper is removed from the anode and deposited on the cathode, leaving the impurities behind.

The chemical changes which occur when a current passes through an electrolyte are called *electrolysis*.

Electrodeposition — an investigation

Use the apparatus described by Fig. 8.1 with copper electrodes and copper sulphate solution. Determine how the amount of copper deposited on the cathode depends on the current used and/or the length of time for which the current is passed through the solution. This will require the cathode to be weighed before and after passing a current through the electrolyte.

A d.c. supply of about 6 V and a current of a few amperes passed for about 10 to 20 minutes will be needed for the amount of copper to be deposited to be of the order of a gramme. The cathode needs to be carefully cleaned, i.e. degreased and rubbed with fine emery, before being used.

Primary cells

A simple cell is produced when two dissimilar electrodes, e.g. one of copper and the other of zinc, are in an electrolyte, see Fig. 8.2. An electric current will flow through an external circuit because a potential difference is produced between the two electrodes.

The size of the e.m.f. produced, i.e. the potential difference between the electrodes when no current is being drawn from the cell (see Chapter 7), depends only on which two metals are used for the electrodes. A table, known as the *electrochemical series*, can be used to determine whether the e.m.f. produced will be large and also which of the materials used will be the positive electrode and which the negative. The following list is part of the electrochemical series, the order of the materials in the list being important:

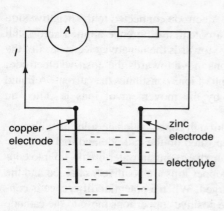

Fig. 8.2 A simple cell

aluminium
zinc
iron
lead
hydrogen
copper
silver
carbon

The further apart the two materials are in the sequence the greater will be the e.m.f. produced and for the two materials selected, the one that is highest in the list will be the negative electrode. Thus for the simple cell described in Fig. 8.2 with electrodes of copper and zinc, because the two metals are quite far apart in the list the e.m.f. produced will be comparatively large and since the zinc is higher than copper in the list it will be the negative electrode.

In the cell, chemical energy is being transformed into electrical energy. With the cell described above the cell ceases to operate when all the appropriate chemicals have been used up. We cannot reverse this change and regenerate the chemicals by supplying electrical energy. Such a cell is called a *primary cell*.

Corrosion

In a domestic-water system there could be copper piping connected to a steel cold-water tank, or other situations may exist where dissimilar metals are in reasonable proximity and there is an electrolyte present such as impure water. The arrangement will then act as a simple cell. This has the effect of gradually destroying the metals by eating away one metal and depositing it on the other. The effect is called *corrosion*.

There are many situations where corrosion due to simple cell action occurs. In some cases it can be because dissimilar metals are close together in a damp atmosphere, in other cases the metal is an alloy and contains within itself dissimilar metals. For instance, brass is an alloy of copper and zinc. Corrosion of this alloy results in a loss of zinc, the process being called *dezincification*.

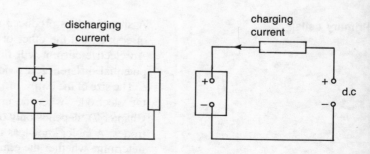

Fig. 8.3 The secondary cell

a) Using the secondary cell. Chemical energy is converted to electrical energy

b) Charging the secondary cell. Electrical energy is converted to chemical energy

Secondary cells

A *secondary cell* is one that can be recharged, i.e. the transformation of chemical energy to electrical energy can be reversed and the cell made usable again, see Fig. 8.3. The lead-acid cell used as the battery in cars is a secondary cell. The battery is used to produce an e.m.f. when the car is being started and then when the car is running it is recharged.

Magnetism, some basic terms

The following are some of the basic terms associated with magnetism.

1 A *permanent magnet* is a piece of material, such as iron, nickel or cobalt, which can be made to be permanently magnetic and does not rely on some external agent to induce magnetism within it.

2 Magnets have two *poles*, called a north and a south pole. These terms are used because a suspended magnet will line up so that its axis lines up in a north—south direction, the end pointing to the north being called the north pole and the other being the south pole. This is the principle of the compass.

3 *Like poles repel, unlike poles attract.* Magnetic forces exist between the poles of magnets. A magnet will attract a piece of iron by temporarily converting it to a magnet so that it has an unlike pole nearest to the magnet.

4 The region in the vicinity of a magnet where a magnetic force is experienced is called a *magnetic field*. The direction of a magnetic field at a point is the direction of the force that would occur if a north pole were placed at the point.

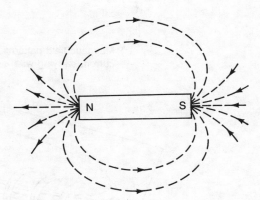

Fig. 8.4 Magnetic field pattern for a bar magnet

5 A magnetic field can be described by *magentic field lines* or, as they are more commonly referred to, *lines of force*. The field lines are drawn so that their directions show the directions of the magnetic field. Figure 8.4 shows the lines of force around a bar magnet. The greater the number of lines of force passing through an area the greater is the strength of the magnetic field.

Magnetic effect of a current

When a current passes through a conductor a magnetic field is produced. Figure 8.5(*a*) shows the field pattern produced round a current-carrying wire, while Fig. 8.5(*b*) shows the field pattern in and around a solenoid carrying a current. The term *solenoid* is used to describe a coil of wire which has a length much greater than its diameter and generally involves many turns of wire. The solenoid gives a magnetic-field pattern very like that of the bar magnet.

The term *electromagnetic* is used for the 'magnet' produced as a result of passing a current through a coil. To increase the strength of this magnet the coil can be wound on a core of soft iron rather than just having air in its core.

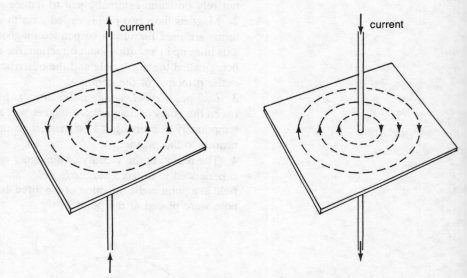

current   current

a) Magnetic field patterns around a long straight
   current carrying wire

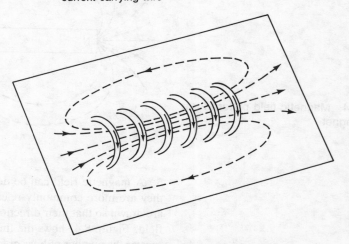

b) Magnetic field pattern due to a current in
   a solenoid

Fig. 8.5

The force on a current-carrying conductor

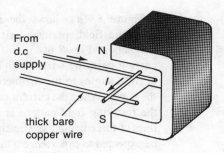

Fig. 8.6 Force on a current-carrying conductor in a magnetic field

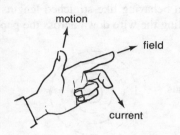

Fig. 8.7 Fleming's left-hand rule

When a current flows through a conductor and it is in a magnetic field which is at right angles to it then the conductor experiences a force. Figure 8.6 illustrates this. The cross piece of wire is free to move. When a current flows through it, and the magnetic field is present, it moves along the two wires. It experiences a force which is at right angles to both the magnetic field and the direction of the current in the conductor.

The direction of the force on a current carrying conductor in a magnetic field can be described by *Fleming's left-hand rule*. If the thumb, the first and second fingers of the left hand are extended so that they are all at right angles to each other, see Fig. 8.7, then the first finger points in the direction of the magnetic field, the second finger in the direction of the current and the thumb in the direction of the motion resulting from the force.

The 'catapult' field

Figure 8.8 shows the patterns of lines of force that occur between two magnets. In *a*) there is attraction between the two magnets whereas in *b*) there is repulsion. Looking at the patterns of the lines of force it can be argued that the lines of force behave rather like stretched lengths of elastic. In *a*) they are trying to pull the magnets together, hence the attraction, while in *b*) they are pulling the magnets away from each other, hence the repulsion.

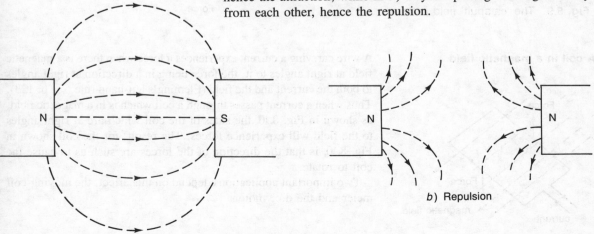

a) Attraction   **Fig. 8.8** Lines of force   b) Repulsion

Figure 8.9(a) shows the field pattern obtained with a uniform magnetic field, perhaps existing in the central region between the magnets in Fig. 8.8(a). In (b) the field pattern is shown due to a current-carrying wire which is at right angles to the paper and hence at right angles to the magnetic field in (a). The current flow is into the paper. When the current-carrying conductor is in the magnetic field the resulting field pattern is as shown in (c). Following the idea of lines of force being like stretched lengths of elastic then the wire would be expected to experience a force moving it downwards across the surface of the paper. This type of field pattern can be called a 'catapult' field in that the lines of force in behaving like stretched lengths of elastic are like a catapult propelling the wire down across the paper.

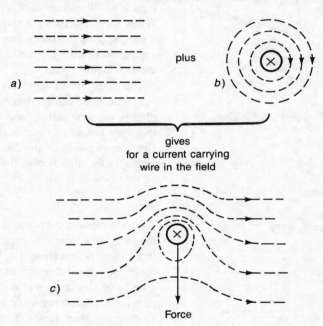

Fig. 8.9 The 'catapult' field pattern

A coil in a magnetic field

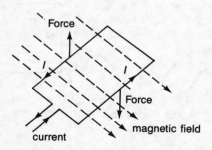

Fig. 8.10 Coil in a magnetic field

A wire carrying a current experiences a force when there is a magnetic field at right angles to it, the force being in a direction at right angles to both the current and the field (Fleming's left-hand rule, see p. 123). Thus when a current passes through a coil which is in a magnetic field, as shown in Fig. 8.10, the sides of the coil which are at right angles to the field will experience forces. The result, for the coil shown in Fig. 8.10, is that the direction of the forces are such as to cause the coil to rotate.

Two important applications depend on this effect, the moving-coil meter and the d.c. motor.

The moving coil meter

Moving-coil ammeters and voltmeters are based on the same principle. This is that a coil carrying a current in a magnetic field experiences forces which cause it to rotate. This basic arrangement is called the moving-coil meter and is shown in Fig. 8.11. The coil is wound on a rectangular frame which is mounted on bearings so that it can rotate freely. Inside the coil is a fixed cylinder of iron. This, together with the shape of the permanent magnet, enables the lines of force to be always at right angles to the sides of the coil, no matter how the coil rotates. The lines of force are said to be radial lines of force in that they are along the lines radiating out from the centre of the iron cylinder.

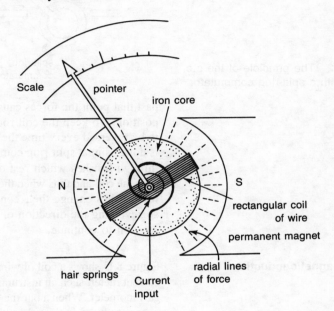

Fig. 8.11   The moving-coil meter

When a current flows through the coil, forces act on the vertical sides of the coil which cause it to rotate. The size of the forces acting on the coil depend on the size of the current, the force being proportional to the current. Opposing the rotation of the coil are hairsprings, one on the top of the coil and one underneath. The coil thus rotates until the force resulting from the current is just balanced out by the force resulting from tightening up the hairsprings. This deflection of the coil shows as a movement of the pointer across a scale and is proportional to the size of the current.

The principle of the d.c. motor

The principle of the d.c. motor can be illustrated by considering a single turn coil as shown in Fig. 8.12. This coil is mounted in a magnetic field. When a current flows through the coil it experiences forces which cause it to rotate. However, the forces would only cause the coil to rotate until it is at right angles to the field. Once it goes

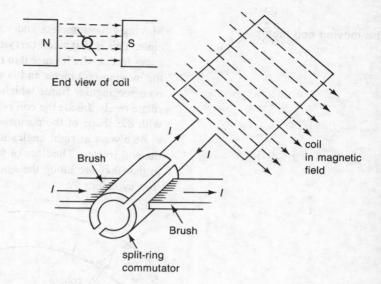

End view of coil

Brush

Brush

split-ring
commutator

coil
in magnetic
field

Fig. 8.12 The principle of the d.c.
motor with a split-ring commutator

past that point the forces cause it to rotate back to the same vertical
position. To keep the coil rotating, the current through the coil has
to be reversed every time the coil reaches the vertical position. This
is done using a split ring commutator. The current to the coil passes
through brushes which rest on the split ring. The split ring rotates
with the coil and thus when the coil passes through the vertical position
the brushes change their contacts from one split ring to the other,
so reversing the direction of current through the coil and enabling
rotation to continue.

Electromagnetic induction

Figure 8.13 shows a coil of wire connected to a sensitive centre-reading
current meter, such an instrument usually being called a centre reading
galvanometer. When a bar magnet is stationary, close to the coil, there
is zero current indicated by the meter. When, however, the magnet
moves towards the coil, current flows in the circuit and a reading is
obtained; the faster the movement of the magnet towards the coil the
bigger the reading. If the magnet is moved away from the meter a
reading is obtained, but a current in the opposite direction to that obtained when the magnet approached the coil. If the magnet is stationary

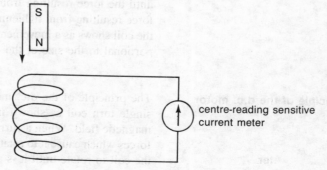

centre-reading sensitive
current meter

Fig. 8.13 Demonstrating electromagnetic induction

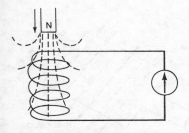

Fig. 8.14   Cutting lines of force

and the coil towards or away from the magnet then readings are again obtained. A current is indicated whenever there is relative movement between the magnet and the coil. The direction of the current depends on whether the magnet and coil are moving closer together or further apart. The size of the current depends on the speed of approach or movement apart. The coil has an e.m.f. induced in it by this relative movement and the higher the speed of approach or movement apart the greater the induced e.m.f. The effect is called *electromagnetic induction*.

Figure 8.14 shows the lines of force associated with the end of the bar magnet as it approaches the coil. When the magnet is moving the lines of force cut through the turns of wire of the coil. When the magnet is stationary relative to the coil there are no lines of force cutting through the turns of wire. It is thus possible to say that an e.m.f. is induced in the coil whenever magnetic lines of force cut through the turns of the coil.

Another way of considering the situation is in terms of magnetic flux. We can imagine there to be a quantity called *flux* which flows along the lines of force. There is an e.m.f. induced when there is a changing amount of flux passing through the coil. The term 'linked' is used for the magnetic flux passing through the coil. The e.m.f. induced is zero when the amount of flux linked by the coil is not changing and a maximum when there is a maximum rate of change of flux linked.

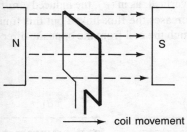

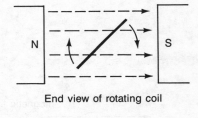

End view of rotating coil

➝ coil movement

a) No change in flux
linked, no induced
e.m.f.

b) Change in flux
linked, e.m.f.
induced

c) Change in flux linked,
e.m.f. induced

Fig. 8.15

Figure 8.15 illustrates some situations involving magnetic flux linked by a coil. You can also consider the situations in terms of the coil cutting lines of force. There are no lines of force cut when there is no change of flux linked by the coil.

The principle of an a.c. generator

Figure 8.16 illustrates the basic principle of an a.c. generator. It consists of a coil of wire that is rotated in a magnetic field. This causes an e.m.f. to be induced in the coil. This e.m.f. gives rise to a current which is fed from the coil through slip rings to brushes which rub against the rings as they rotate. The output is an alternating e.m.f.

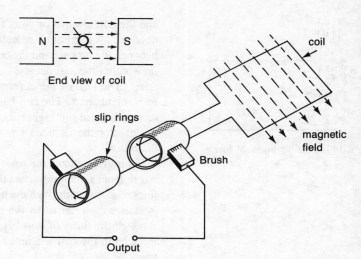

End view of coil

slip rings

Brush

Output

Fig. 8.16 The principle of an a.c. generator

coil

magnetic field

Figure 8.17 illustrates how the flux linked by the coil changes as it rotates. With the coil at right angles to the magnetic field, as in (a), a slight rotation of the coil barely changes the flux linked by the coil and thus there is no e.m.f. induced. As the coil further rotates the flux linked by the coil changes, reaching a maximum rate of change when the coil is in position (b). This then results in a maximum e.m.f. at that time. Further rotation reduces the flux linked and thus the induced e.m.f. until when the coil is again vertical, as in (c), the induced e.m.f. is again zero. Further rotation increases the flux linked but this time the flux is effectively passing through the coil from the opposite direc-

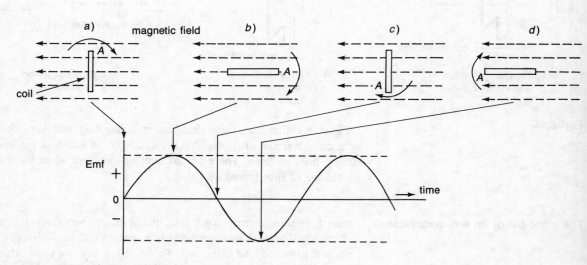

Fig. 8.17 The output from a coil rotating in a magnetic field

tion. The result is an induced e.m.f. but in the opposite direction. This e.m.f. reaches a maximum when the coil has rotated to position (*d*). The flux linked then reduces until position (*a*) is reached again. Then the entire cycle repeats itself.

elf-assessment questions

1 How much energy in joules is used by a 2 kW electric fire if it is on for 4 hours?

2 How many joules are there in a one kilowatt hour?

3 In a household during a day a 500 W TV set is used for 5 hours, four 100 W lamps for an average of 4 hours each, the central-heating boiler which uses 500 W for a total time of 12 hours and a 2 kW electric fire for 2 hours. How many kilowatt hours of electricity are used in the day?

4 Explain the following terms: *a*) electrode, *b*) electrolyte, *c*) electrolysis, *d*) anode, *e*) cathode, *f*) ion.

5 Explain how, when electrolysis occurs with copper sulphate and copper electrodes, the cathode increases in weight while the anode decreases.

6 Which of the following cells would you expect to give the largest e.m.f. and in each case which electrode would be the positive terminal?
a) aluminium — copper
b) iron — copper
c) copper — silver

7 Explain the terms primary cell and secondary cell.

8 Explain the terms magnetic field and magnetic lines of force.

9 An electromagnet is frequently used with a crane for lifting steel and iron in a scrap metal yard. Explain why an electromagnet is preferred to a permanent magnet or even just a hook.

10 Figure 8.18 shows an electric-bell circuit. Explain what happens when somebody presses the bell push and so closes the switch *S*.

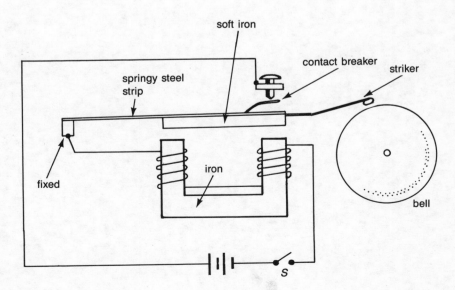

Fig. 8.18 An electric bell

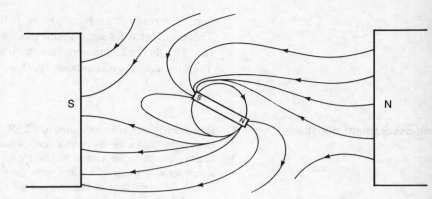

S

N

Fig. 8.19

11 Figure 8.19 shows the lines of force pattern occurring when a small permanent magnet is put in the magnetic field of a larger magnet. If the smaller magnet can move, explain what you would expect to happen. .

12 Explain the principle of operation of the moving-coil meter.

13 Explain the principle of operation of the d.c. motor.

14 Explain what is meant by electromagnetic induction and describe a simple experiment which could be used to demonstrate it.

15 Explain the principle of the a.c. generator.

Guide to answers to self-assessment questions

The following are intended as brief guides as to how the answers can be obtained to the problems at the ends of each chapter. They are not intended to be model answers but purely aids to enable you to work out the answers.

Chapter 1

1. *a*) kg, *b*) m, *c*) s, *d*) A, *e*) K.
2. *a*) 12 mm, *b*) 2.56 km, *c*) 0.120 m, *d*) 1.2 mA, *e*) 1.60×10^{-4} A or 0.000 160 A, *f*) 4.32 kA.
3. (m/s) × (s) = m.
4. $(\text{kg m/s}^2)/(\text{m}^2) = \text{kg}/(\text{m s}^2)$.
5. *a*) Straight line through the origin. *b*) Slope about 2.4 g/mm, hence load = 2.4 × extension, when load is in g and extension is in mm.
6. *a*) Straight line through origin. *b*) Slope about 4.1 ohm/cm, hence resistance = 4.1 × length, when resistance in ohms and length in cm.
7. *a*) A straight line not passing through the origin. *b*) Slope about 0.20 ohm/°C, intercept on resistance axis about 21.0 ohm, hence resistance = 0.20 × temperature + 21.0.
8. *a*) Straight line through origin, slope *R*. *b*) Straight line through origin, slope *k*. *c*) Not straight line. *d*) Straight line passing through origin, slope *L*α. *e*) Not a straight line. *f*) A straight line not passing through the origin, slope *a* and intercept on *v* axis of *u*.

Chapter 2

1. See text.
2. See text.
3. *a*) 1 copper, 1 oxygen. *b*) 2 hydrogen, one oxygen. *c*) 3 iron, 4 oxygen. *d*) 1 carbon, two oxygen.
4. See text.
5. Electrons orbiting a nucleus.

6. 78% nitrogen, 21% oxygen, 0.93% argon, 0.03% carbon dioxide, when dry.
7. Compounds formed during a reaction between an element and oxygen.
8. It combines with the oxygen to produce an oxide.
9. Oxygen in the air combines with the copper to give copper oxide. The extra weight is due to the oxygen.
10. Oxygen (air) and water.
11. Coating with paint, grease, plastic film or a more corrosion resistant metal.
12. Bodywork of a car, steel girders of bridges, iron railings, etc.
13. a) $x = 1$, $y = 2$, b) $x = 1$, $y = 1$, c) $x = 2$, $y = 2$.

Chapter 3

1. See text.
2. Tie is in tension, strut in compression.
3. See text, examples include a spiral spring with low loads.
4. Straight line passing through the origin.
5. 0.4, 0.6, 0.8.
6. $1\frac{1}{2}$ times force means $1\frac{1}{2}$ times extension, i.e. 1.35 mm.
7. 20 N gives extension of 30 mm, $1\frac{1}{2}$ times load means $1\frac{1}{2}$ times extension, i.e. 45 mm. Hence length = 295 mm.
8. $(2.0/1.2) \times 200 = 333$ N.
9. See text.
10. a) 180 N at 34° to 150 N force. b) 68 N at 8° to 50 N force. c) 125 N at 44° to 40 N force.
11. See text.
12. 2.6 kN at 19° to 2.0 kN force.
13. 49 kN in the direction of motion of the ship.
14. a) 43 N and 25 N. b) 230 N and 193 N. c) 2.97 kN and 2.97 kN. d) 2.1 kN and 5.6 kN.
15. Down incline $2.0 \times \sin 15° = 0.52$ kN, at right angles $2.0 \times \cos 15° = 1.93$ kN.
16. Horizontal $10 \times \cos 40° = 7.7$ kN, vertical $10 \sin 40° = 6.4$ kN.
17. See text.
18. See text.
19. 45 N at about 153° (i.e. 180° −27°) from 40 N force.
20. a) 64.7 N at about 154° to 30 N force. b) 659 N at about 163° to 400 N force. c) 1.6 kN at about 68° to 2.0 kN force. d) 392 N at about 107° to 250 N force.
21. Force = $100 \times 9.8 = 980$ N, hence tension = 693 N (same in both ropes).
22. See text.
23. a) $20 \times 500 = 10\ 000$ N mm, b) $2.0 \times 1.5 = 3.0$ kN m, c) $50 \times 10 = 500$ N cm, d) $30 \times 12 = 360$ N mm.
24. $800 \times d = 600 \times 2.0$, hence $d = 1.5$ m.
25. a) $F \times 50 = 100 \times 20$, hence $F = 40$ N. b) $F \times 500 = 300 \times 100$, hence $F = 60$ N. c) $F \times 100 = 5g \times 150 + 12g \times$

200, hence $F = 31.5g = 31.5 \times 9.8 = 309$ N or the weight of a mass of 31.5 kg. *d*) $400g \times 10 = F \times 12 + 100g \times 32$ and so $F = 66.7g = 66.7 \times 9.8 = 653$ N or the weight of a mass of 66.7 kg.

26. *a*) $200 \times 50 = R_2 \times 100$, hence $R_2 = 100$ N. $R_1 + R_2 = 200$, hence $R_1 = 100$ N. *b*) $3.0 \times 1.0 + 1.0 \times 3.8 = R_2 \times 1.8$, hence $R_2 = 3.8$ N. $R_1 + R_2 = 3 + 1$, hence $R_1 = 0.2$ N. *c*) $50 \times 0.2 + 30 \times 0.5 = R_2 \times 1.0$, hence $R_2 = 25$ N. $R_1 + R_2 = 50 + 30$, hence $R_1 = 55$ N. *d*) $500 \times 0.2 + R_2 \times 0.3 = 200 \times 0.5$, hence $R_2 = 0$ N. $R_1 + R_2 = 500 + 200$, hence $R_1 = 700$ N.

27. See text.

28. *a*) $R_1 = R_2 = 3.0 \times 9.8 = 29.4$ N. *b*) $R_1 = R_2 = 38.0 \times 9.8 = 372.4$ N.

29. See text.

Chapter 4

1. *a*) $p = h\rho g = 1.5 \times 1000 \times 9.8 = 14\ 700$ Pa. *b*) $p = F/A$, hence $F = p \times A = 14\ 700 \times 4.0 \times 2.0 = 117\ 600$ N.

2. $p = h\rho g = 1.2 \times 0.8 \times 1000 \times 9.8 = 9408$ Pa.

3. *a*) $p = h\rho g = 4.0 \times 0.70 \times 1000 \times 9.8 = 27\ 440$ Pa. *b*) Half depth of petrol so half pressure, i.e. 13 720 Pa.

4. $p = h\rho g = (120/1000) \times 1000 \times 9.8 = 1176$ Pa.

5. $p = h\rho g = (750/1000) \times 13\ 600 \times 9.8 = 99\ 960$ Pa.

6. $p = h\rho g$, hence $h = 100 \times 1000/(1000 \times 9.8) = 10.2$ m.

7. $p = h\rho g = (120/1000) \times 1000 \times 9.8 = 1176$ Pa. This is the gauge pressure and it is higher.

8. Absolute pressure $= 20 + 100 = 120$ kPa.

9. $F_2 = (A_2/A_1) \times F_1$, thus $1400 = (1600/200) \times F_1$ and $F_1 = 175$ N.

10. The pressure increases with depth.

11. *a*) Volume $= (80/1000)/7000 = 1.1 \times 10^{-5}$ m^3. Upthrust $= 1.1 \times 10^{-5} \times 1000 \times 9.8 = 0.11$ N. Weight $= (80/1000) \times 9.8 = 0.78$ N. It sinks. *b*) Volume $= (10/1000)/250 = 4.0 \times 10^{-5}$ m^3. Upthrust $= 4.0 \times 10^{-5} \times 1000 \times 9.8 = 0.39$ N. Weight $= (10/1000) \times 9.8 = 0.098$ N. It rises. *c*) Upthrust $= (10/1\ 000\ 000) \times 1000 \times 9.8 = 0.098$ N. Mass $= 920 \times (10/1\ 000\ 000) = 0.092$ kg and so weight $= 0.092 \times 9.8 = 0.090$ N. It rises. *d*) Upthrust $= 15 \times 1.2 \times 9.8 = 176$ N. Weight $= 10 \times 9.8 = 98$ N. It rises.

12. Upthrust $=$ weight $= 300 \times 1000 \times 9.8 = 2.94 \times 10^6$ N.

Chapter 5

1. Speed $=$ distance/time $= 8.0 \times 1000/(80 \times 60) = 1.7$ m/s.

2. Distance $=$ speed $\times$ time $= 60 \times (20/60) = 20$ km.

3. Distance $= 200$ m. Displacement $= \sqrt{(100^2 + 100^2)} = 141$ m NW.

4. Vectors have size and direction, scalars only have size.
5. Average speed = total distance/time = 2/6 = 0.33 km/min.
6. Distance = speed × time, hence in 20 minutes, distance = 60 × (20/60) = 20 km and in the 10 minutes, distance = 80 × (10/60) =13 km. Hence total distance = 33 km.

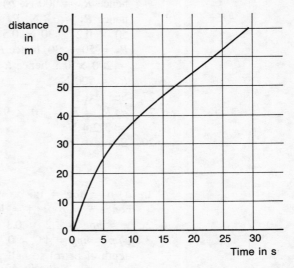

Fig. A1

7. See Fig. A1. *a*) (i) constant speed, (ii) accelerating, (iii) constant speed. *b*) Slope = speed. (i) 5 m/s, (ii) 1.7 m/s.

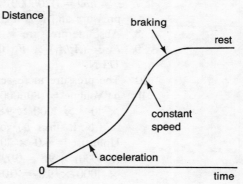

Fig. A2

8. See Fig. A2.
9. *a*) Speed = 2 m/s. *b*) Speed = 0 m/s. *c*) Speed = (40−20)/(25−15) = 2 m/s. *d*) Speed = (40−0)/(25−0) = 1.6 m/s.
10. Average acceleration = change in velocity/time = (20 × 1000/3600)/12 = 0.46 m/s^2.
11. Average acceleration = change in velocity/time = (20 × 1000/3600)/2 = 2.8 m/s^2.
12 Average acceleration = change in velocity/time = (0−40)(1000/3600)/5.0 = −2.2 m/s^2.

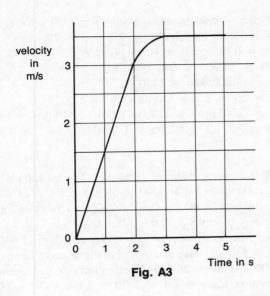

Fig. A3

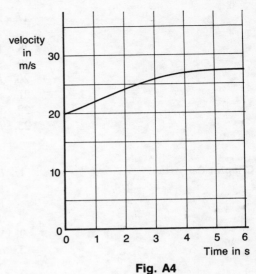

Fig. A4

13. *a*) See Fig. A3. *b*) (i) Accelerating, (ii) constant velocity. *c*) Slope = 1.5 m/s^2 *d*) Area = 13 m.
14. *a*) See Fig. A4. *b*) (i) Accelerating, (ii) constant velocity. *c*) (i) Slope = 2 m/s^2, (ii) Slope = 0.15 m/s^2. *d*) (i) Area = 22.5 m, (ii) area = 153 m.

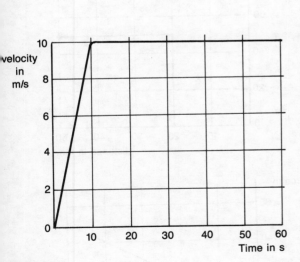

Fig. A5

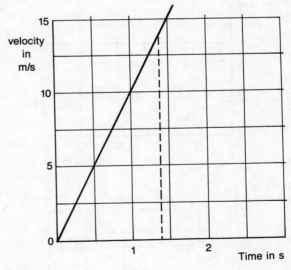

Fig. A6

15. See Fig. A5.
16. Slope of the graph must be 9.8 m/s^2 and so velocity changes by 9.8 m/s for each second. See Fig. A6. The area under the graph must be 10 m. This is about a time of 1.4 s and a velocity of 14 m/s.

17. Acceleration = velocity change/time, hence *a*) velocity = 9.8 × 2 = 19.6 m/s, *b*) velocity = 9.8 × 3 = 29.4 m/s.

18. $F = \mu N = 0.70 \times 6.0 \times 9.8 = 41$ N.

19. $\mu = F/N = 100/(15 \times 9.8) = 0.68$.

20. $N = F/\mu = 1000/0.80 = 1250$ N.

21. $v = f\lambda = 50 \times 3.0 = 150$ mm/s.

22. $v = f\lambda = (200/1000) \times (1.7 \times 1000) = 340$ m/s.

23. $f = v/\lambda = 300\,000\,000/200 = 1.5$ MHz.

Chapter 6

1. *a*) Work = force × distance = 50 × 1.2 = 60 J. *b*) Work = force × distance = 1.0 × 1000 × 50/100 = 500 J. *c*) Work = force × distance = 600 × 0.40 = 240 J. *d*) Work = force × distance = 4.0 × 9.8 × 2.0 = 78.4 J.

2. Work = force × distance = 30 × 9.8 × 60 = 17 640 J.

3. *a*) Work = force × distance moved in direction of force = 100 × (30/100) cos 30° = 26 J. *b*) Work = force × distance moved in direction of force = 1.4 × 1000 × 1.1 cos 45° = 1089 J.

4. Work = force component in direction of motion × distance = 1000 × 9.8 × sin 10° × 30 = 51 053 J.

5. See Fig. A7. Work done = area under graph. *b*) 50 × 50/1000 = 2.5 J. *b*) (20 × 20/1000) + (50 × 30/1000) = 1.9 J. *c*) ½ ×

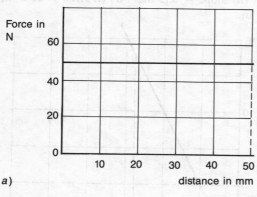

a)

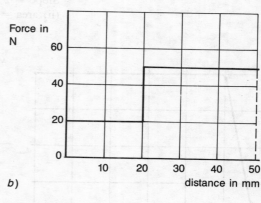

b)

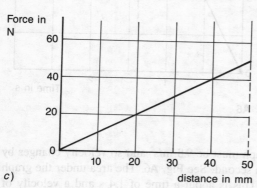

c)

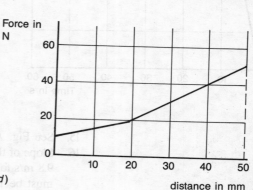

d)

Fig. A7

$50 \times 50/1000 = 1.25$ J. *d*) $(10 \times 50/1000) + (\frac{1}{2} \times 10 \times 20/1000) + (10 \times 30/1000) + (\frac{1}{2} \times 30 \times 30/1000) = 1.35$ J.

6. *a*) Energy associated with position, *b*) energy associated with motion.

7. *a*) Work = force × distance = $500 \times 0.60 = 300$ J. *b*) Power = work/time = $300/2.0 = 150$ W.

8. Work = force × distance = $50 \times 9.8 \times 4.0$ J, power = work/time = $50 \times 9.8 \times 4.0/60 = 32.7$ W.

9. Power = work/time = force × distance/time = $20 \times 40 \times 9.8 \times 4.0/(2.0 \times 60) = 261$ W.

10. Power = force × velocity, hence force = $20 \times 1000/(50 \times 1000/3600) = 1440$ N.

11. *a*) Efficiency = (useful output energy/input energy) × 100, hence if we consider energy in 1 s, then output energy in 1 s = $2.0 \times 60/100 = 1.2$ kJ. Output power = 1.2 kW. *b*) Power = work/time, hence work = $1.2 \times 2.0 \times 60 = 144$ kJ.

12. *a*) Work = force × distance = $500 \times 6.0 = 3\ 000$ J. *b*) Output power = work/time = $3000/10 = 300$ W. *c*) Efficiency = (useful output energy/input energy) × 100, hence input power = $300/(70/100) = 429$ W.

13. Heat capacity = Q/t, hence $Q = 2500 \times 200 = 500\ 000$ J.

14. Specific heat capacity = Q/mt, hence $Q = (500/1000) \times 4200 \times (100-20) = 168\ 000$ J.

15. Specific heat capacity = Q/mt, hence $Q = 12 \times 500 \times 280 = 1\ 680\ 000$ J

16. Sensible heat results in a temperature change, latent heat gives no such change in temperature but a change of state.

17. Specific latent heat = Q/m, hence $Q = 12 \times 335 = 4020$ kJ.

18. To heat water from 20 °C to 100 °C, $Q = m \times c \times t = 2.0 \times 4200 \times 80 = 672\ 000$ J. To change water at 100 °C to steam at 100 °C, $Q = mL = 2.0 \times 2257 \times 1000 = 4\ 514\ 000$ J. Hence heat required = $5\ 186\ 000$ J.

19. To heat iron from 20 °C to 1200 °C, $Q = m \times c \times t = 500 \times 0.50 \times 1180 = 295\ 000$ kJ. To melt iron $Q = mL = 500 \times 270 = 135\ 000$ kJ. Hence heat required = $430\ 000$ kJ.

20. Linear expansivity = change in length/(original length × change in temperature). Hence change in length = linear expansivity × original length × change in temperature. For concrete, change in length = $0.000\ 014 \times 100 \times 20 = 0.028$ m. For steel, change in length = $0.000\ 011 \times 100 \times 20 = 0.022$ m.

21. Linear expansivity = change in length/(original length × change in temperature). Hence change in length = linear expansivity × original length × change in temperature = $0.000\ 017 \times 100 \times 5 = 0.0085$ m. Hence length = 100.0085 m.

22. Linear expansivity = change in length/(original length × change in temperature), hence change in length = linear expansivity × original length × change in temperature) = $0.000\ 017 \times 7.0 \times 55 = 0.0065$ m.

23. Apparent cubic expansivity = apparent change in volume/ (original volume × change in temperature), hence apparent change in volume = apparent cubic expansivity × original volume × change in temperature = 0.000 18 × 1000 × 55 = 9.9 cm³. This is the amount overflowing.

24. Cubic expansivity = apparent cubic expansivity + cubic expansivity of container. Hence apparent cubic expansivity = 0.001 20 − (3 × 0.000 011) = 0.001 17/K. Apparent change in volume = apparent cubic expansivity × original volume × change in temperature = 0.001 17 × 4.0 × 10 = 0.046 8 m³. This is the amount that will overflow.

25. See Fig. 6.12. When the temperature rises the strip curves and at the selected temperature will break the electrical contact. When the temperature falls to the selected temperature the strip has straightened sufficiently to make the electrical contact and switch on the heater.

Chapter 7

1. *a*) ampere, *b*) volt, *c*) watt, *d*) ohm.
2. See Fig. 7.4. d.c. gives a current always flowing in one direction, a.c. gives a current which periodically alternates its direction.
3. Average current = charge/time, hence charge = current × time = 2 × 1 = 2 C.
4. Average current = charge/time, hence charge = current × time = 4 × 5 × 60 = 1200 C.
5. $R = V/I = 10/0.50 = 20\ \Omega$.
6. $R = V/I$, hence $V = IR = (120/1000) \times 20 = 2.4$ V.
7. $R = V/I$, hence $I = V/R = 6.0/12 = 0.50$ A.
8. *a*) See text. *b*) Straight line passing through the origin. *c*) The potential difference will be doubled.
9. *a*) For A, $R = V/I$ or about 3.2/1.Q = 3.2 Ω. For B, $R = V/I$ or about 0.75/1.0 = 0.75 Ω. *b*) For A, $R = V/I$ or about 6.0/3.0 = 2.0 Ω. For B, $R = V/I$ or about 6.0/3.0 = 2.0 Ω.
10. *a*) Both. *b*) See Fig. A8. The ratio resistance/p.d. is a constant and so the current does not change. If you are not sure put scales on the graph and calculate some currents.
11. *a*) In series. *b*) In parallel.
12. *a*) Yes. *b*) No.
13. *a*) No. *b*) Yes.
14. *a*) $R = R_1 + R_2 = 10 + 20 = 30\ \Omega$. *b*) $1/R\ 1/R_1 + 1/R_2$, hence $R = 6.7\ \Omega$. *c*) $R = R_1 + R_2 + R_3 = 2 + 4 + 6 = 12\ \Omega$.
15. *a*) $1/R = 1/20 + 1/50$, hence $R = 14.8\ \Omega$. With 10 Ω this is 24.3 Ω. *b*) $1/R = 1/2 + 1/1$, $R = 0.67\ \Omega$. *c*) $1/R = 1/10 + 1/5$, hence $R = 3.3\ \Omega$ and with 15 Ω is 18.3 Ω. *d*) $1/R_1 = 1/2 + 1/6$, hence $R_1 = 1.5\ \Omega$. $1/R_2 = 1/8 + 1/12$, hence $R_2 = 4.8\ \Omega$. $R_1 + R_2 = 6.3\ \Omega$.
16. *a*) $R = V/I$, hence $I = V/R = 12/10 = 1.2$ A. *b*) 12 V.
17. *a*) $R = 10 + 15 = 25\ \Omega$, hence $I = V/R = 12/25 = 0.48$ A. *b*) $V = IR = 0.48 \times 15 = 7.2$ V.

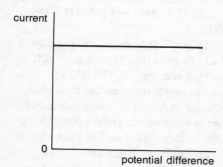

current

0

potential difference

Fig. A8

18. $1/R = 1/5 + 1/3$, hence $R = 1.9\ \Omega$. Total resistance $= 2 + 1.9$ $= 3.9\ \Omega$. $I = V/R = 2/3.9 = 0.51$ A. Potential difference across $R, V = IR = 0.51 \times 1.9 = 0.97$ V. Hence $I_1 = V/5 = 0.97/5$ $= 0.19$ A. $I = I_1 + I_2$, hence $I_2 = 0.51 - 0.19 = 0.32$ A.

19. $1/R = 1/6 + 1/12$, hence $R = 4\ \Omega$. $I = V/R = 6/4 = 1.5$ A. $I_1 = 6/6 = 1.0$ A. $I = I_1 + I_2$, hence $I_2 = 0.5$ A.

20. $1/R_1 = 1/6 + 1/3$, hence $R_1 = 2\ \Omega$. Total resistance $= R_1 + R$ $= 2 + R$. Total resistance $= V/I = 12/2.0 = 6.0\ \Omega$. Hence $2 + R = 6$ and $R = 4\ \Omega$.

21. $1/R = 1/20 + 1/10$, hence $R = 6.7\ \Omega$. Total resistance $= 16.7\ \Omega$. $V = I \times$ total resistance $= 1.5 \times 16.7 = 25$ V.

22. *a*) $V = 3 + 5 + 8 = 16$ V. *b*) $R = V/I = 16/2 = 8\ \Omega$. *c*) $R = V/I = 3/2 = 1.5\ \Omega$, $R = 5/2 = 5.2\ \Omega$, $R = 8/2 = 4.0\ \Omega$.

23. *a*) Power $= VI$, hence $I = 100/240 = 0.42$ A. *b*) $R = V/I = 240/0.42 = 571\ \Omega$. *c*) Power $=$ energy/second, hence in 1 hour energy $= 100 \times 60 \times 60 = 360\ 000$ J.

24. Power $= I^2R = 4.0^2 \times 5.0 = 80$ W.

25. Power $= V^2/R = 6^2/50 = 0.72$ W.

26. Power $= I^2R = 10^2 \times 2 = 2500$ W.

27. Power $= VI$, hence $I =$ power/V. *a*) $I = 120/240 = 0.5$ A. 2 A fuse. *b*) $I = 2000/240 = 8.3$ A. 10 A fuse. *c*) $I = 500/240 = 2.1$ A. 5 A fuse.

Chapter 8

1. $2000 \times 4 \times 60 \times 60 = 28\ 800\ 000$ J.

2. See text, 3 600 000 J.

3. $0.500 \times 5 + 0.100 \times 4 \times \propto \Rightarrow 4 + 0.500 \times 12 + 2 \times 2 = 14.1$ kilowatt hours.

4. See text.

5. See text.

6. The aluminium-copper cell. Positive terminals *a*) copper, *b*) copper, *c*) silver.

7. See text.

8. See text.

9. An electromagnet can be switched off and so drop the load, a permanent magnet can not. A magnet attracts the metal without needing it to be attached to a hook.

10. When the switch is closed a current passes through the coil and it acts as a magnet and attracts the piece of soft iron. This causes the striker to hit the bell. However, the movement of the soft iron causes the electric circuit to be broken and so the coil no longer acts as a magnet and the striker moves away from the bell. The circuit is then made again and the movement is repeated.

11. The small magnet rotates to line up with its south pole pointing to the north pole of the large magnet and its south pole to the north pole. It is effectively a small compass needle.

12. See text, Fig. 8.11.
13. See text, Fig. 8.12.
14. See text, Fig. 8.13.
15. See text, Fig. 8.16.

Index